The Roman Milestones of Britain:

their Petrography and probable Origin

Jeffrey P. Sedgley

British Archaeological Reports 18
1975

British Archaeological Reports

122 Banbury Road, Oxford OX2 7BP, England

B.A.R. 18, 1975, "The Roman Milestones of Britain".
© Jeffrey P. Sedgley, 1975.

ISBN 9780904531206 paperback
ISBN 9781407318097 e-book
DOI   https://doi.org/10.30861/9780904531206
A catalogue record for this book is available from the
British Library
This book is available at www.barpublishing.com

THE MILESTONES OF ROMAN BRITAIN

LIST OF ILLUSTRATIONS

Page

FIGURES

ACKNOWLEDGEMENTS

This article is based on a thesis submitted for the degree of M.Sc. at the University of Keele, 1969. I would like to thank Professor F.W. Cope for making funds available for this research and for useful discussion with him and other members of the Department of Geology. Also I am grateful to the various museums, organisations and individuals who allowed me to take samples from their milestones; without such co-operation this survey would not have been possible. Professor S.S. Frere kindly made useful criticisms of an earlier draft of this paper. Finally I would like to thank Professor A.L.F. Rivet, whose idea this research was, for much helpful guidance, discussion and encouragement on many points.

CONTENTS

1. THE PURPOSE OF THE INVESTIGATION

The purpose of this survey was to determine so far as was possible the original quarry sources of the Roman milestones of Britain. There was reason to suppose that some considerable trouble was taken to secure good stone for milestones: for example, a stone of Constantine from Bourg-St.-Pierre on the Great St. Bernard Pass[1] may have been quarried as far away as the Jura, 75 miles to the north[2], while preliminary studies in Britain suggested that the Hadrianic stone from Rhiwiau-uchaf, Caernarvonshire[3] (51)[4] and probably others may not have been local. However, it seems that by the third century, in Britain at least, almost any local stone was utilised, provided that it could be inscribed.

Before the end of the last century Mommsen[5] published his discussion of the milestones of North Africa. These milestones generally have fairly full inscriptions which usually include a mileage figure and often also name the authority responsible for building and maintaining the roads and for the erection of milestones. In addition Mommsen was able to make use of the case endings of the inscriptions. He was able to point out that in civil areas of the empire roads were basically of two types; public roads of the Roman people, and the roads of the coloniae and municipia and other civitates. The former were mostly built by military labour at the expense of the aerarium and numbered their miles from the capital of the province to its boundaries; the latter were laid through individual territories at the expense of the cites and numbered their miles from the individual towns to the boundaries of their territories. Mommsen thought it probable, however, that in the third and fourth centuries this system was breaking down and even the public roads were repaired at the expense of the municipia, or territory by territory, rather than at the public expense of the Roman people.

Mommsen was well aware that this rule, of the two types of road, could not be followed rigidly for all areas of the empire and the milestones of Roman Britain are poor indeed by comparison with those of North Africa. Nevertheless, Mr. C.E. Stevens has shown how their evidence can be used to elucidate some of the problems of the civitates, their status and territorial extent[6]. If, then, Mommsen's analysis is substantially correct one might expect different patterns of quarry sources to apply to milestones on roads whose maintenance was a central responsibility, and those for which upkeep devolved on the local authority (civitas). In the latter case, bearing in mind the local availability of stone, one would expect the quarries to be within the area of the civitas in question. It was, therefore, hoped that besides throwing some light on the variation of practice in different places and different periods in Roman Britain, the survey might add something to the knowledge of the boundaries between the civitates.

1

2.   THE MILESTONES;   THEIR DISTRIBUTION IN SPACE AND TIME

There are 110 milestones and probable milestones known from Britain.   Twenty-five are noted in literature but have since been lost.   Of the extant stones, 101 inscribed examples are listed in R.I.B. and J.R.S.   Nine uninscribed milestones are found in a variety of sources.

The Hadrianic stone from Thurmaston, Leicester (32) is typical of the second century Romano-British milestones, recording the emperor's title and dates, caput viae (place from which road measurements were made) and mileage figure.   During the third century the milestones often seem to have become little more than vehicles of political propaganda;   e.g. the six inscribed during the eight months of Constantine's caesarship.   They seldom have a town or a mileage figure inscised upon them, though it is possible that the mileage was painted on at the end of the inscription. Only one milestone (90) has been found which had the fourth or fifth century honorific inscription 'B.R.P.N.'.

Milestones are found widespread throughout the province, from Helston in Cornwall to Ingliston near Edinburgh.   (See Fig. 1).   There are more remaining in the north and west of the country, for in the south and east, where building stone is less easy to obtain, they have been more vulnerable to re-use.   The seven recorded stones from Bitterne, for example, were utilised in the late Roman town wall[7].   Only in the region of Hadrian's Wall can the concentration of milestones be said to be high.

Much of the main road system of Roman Britain must undoubtedly have had its foundations in the early years of the invasion, yet the earliest known milestones are three of Hadrian, from Leicester (32), North Wales (51) and Lancaster (61).   The reason for this could be that before Hadrian's time, in most parts of the province at least, the roads were not properly equipped with milestones; the correction of this could have been one of the "many reforms" which Hadrian is said to have instituted in Britain[8].

Except for two stones of Antoninus Pius from Scotland (108 and 109) there are no further milestones recorded until the first 25 years of the third century when there are four.   Why the absence of milestones in England and Wales for such a long time?   Surely this can only be because few, if any, milestones were erected during this period, little road building or repair being necessary. After Septimius Severus had defeated Albinus he strengthened and reorganised the military areas of the country. Concomitant with this would probably have been road rebuilding in the civil areas of the province and this seems to be testified by a milestone from Bitterne (12) which proclaims that the emperor "in the eighteenth year of his tribunician power restored the roads which had fallen into

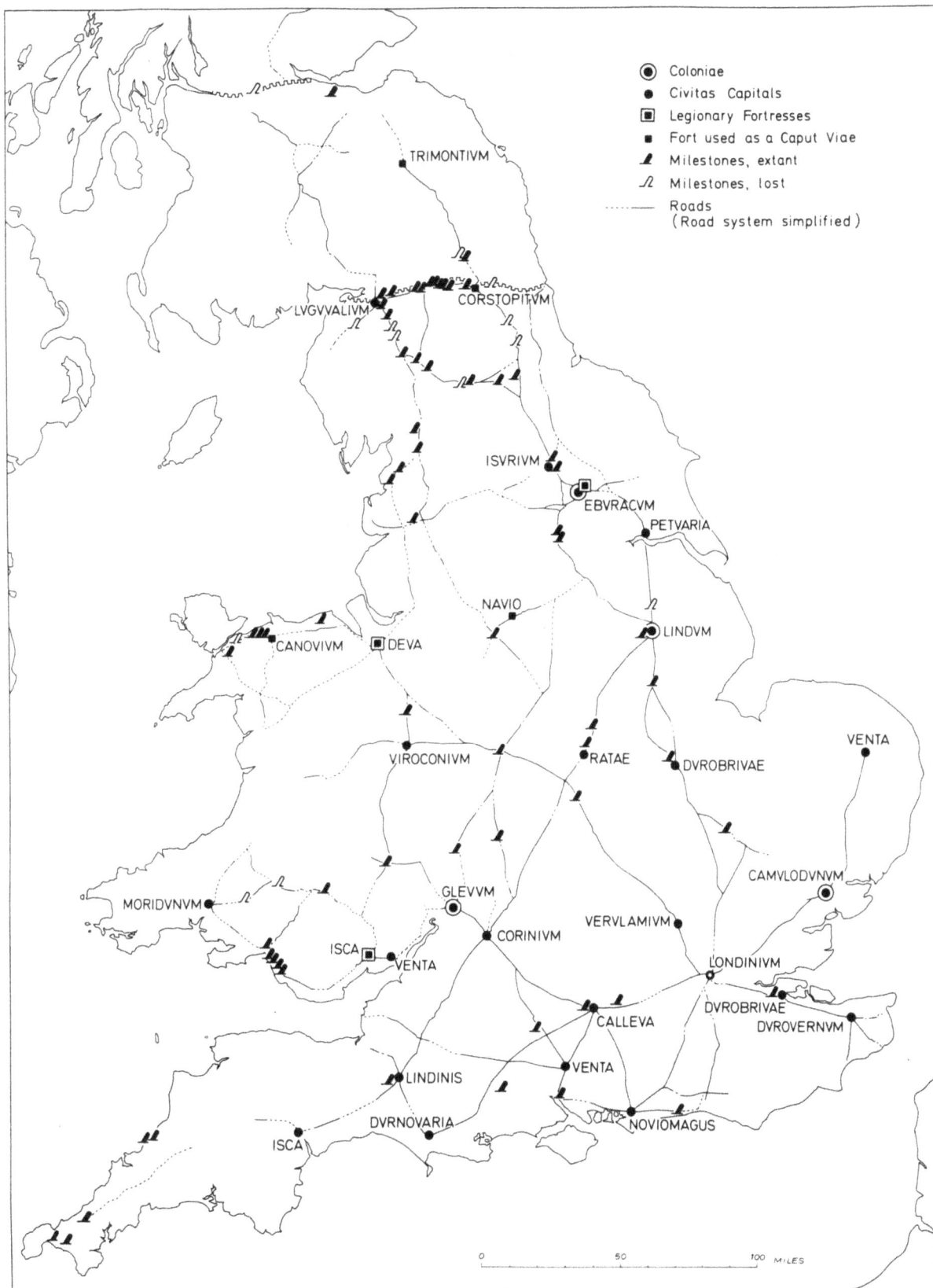

FIG. 1   The distribution of Roman milestones in
Britain. (Drawn by G. Barber)

ruin through age..." It is not until after the reign of Gordian III that the number of milestones begins to increase and thereafter hardly a reign is left unrecorded, the milestones averaging nearly one per year, until that of Crispus which forms the last of the series.

The shape of some stones and the lay-out of their inscriptions suggests that they have been cut down from larger stones. The Constantinian stone from Cambridge (22) may once have been cylindrical for, though it is now roughly quadrangular, it has more or less straight sides but rounded edges. Other stones were turned round (44) or up-ended (85) and re-used. Not infrequently, in areas where stone is plentiful, a new milestone would be erected with the old one remaining on the spot (e.g. 59 and 60); while at Crindledykes on the Stanegate as many as seven stones were found at one spot (92-8), though duplication of some dates suggests that they might have been brought there from two original sites.

Milestones have been put to a variety of secondary uses, the most obvious being as building stone. A stone of Decius (13) was found face down being used as a step in a villa, probably even before it could be erected on a road. This was perhaps considered fitting treatment for Decius as another of his stones was up-ended and re-inscribed (64). Another use is as a gatepost (20) or even as part of a stile (18). One milestone (44) is reputed to have had a trip round a bay as ballast in a pilot boat. Some of the Welsh milestones have been later used as Christian memorial stones (43 and 53)[9].

3.   THE METHOD OF INVESTIGATION

There had been no detailed petrographic work of this type carried out on milestones, but most of the entries in R.I.B. have been given a geological description such as "sandstone", "granite" or "oolite limestone". While these provide a beginning in some instances, often they were described as "gritstone", a term too vague to be of much use, and occasionally a stone was given a completely wrong identification (e.g. 42).

The milestones were sampled by taking a core, 2 cm in diameter and about 5 cm long, with a diamond bit drill. On most museum specimens the bit was attached to a two-speed, heavy duty, electric hand-drill, water being applied to the bit at intervals. For outdoor stones and when electricity was not available, a two-stroke drill was used which had its own pressure water supply. On free standing stones it was often possible to take the sample from the base but all too frequently stones were fixed so that the sample had to be taken from the top, back or side. On one occasion the face of a museum stone had to be drilled because of the way it was fixed in its case. In each instance the resulting hole was filled and suitably disguised. Initially it was intended to plug the hole with the outer part of the core set in polyfilla, leaving only a thin annulus to be painted. It was soon found, however, that the hole could be rendered more nearly invisible by filling it almost completely with polyfilla and then adding a final layer of a mixture of paint, polyfilla and rock chippings. Some of the outdoor stones were later refilled with cement and rock chippings when the polyfilla was found to have contracted and weathered away.

The core obtained was examined in hand specimen and thin section. Where more than one example of a particular sandstone occurred, as with the Carboniferous Yoredale sandstones and the St. Bees Sandstone, more detailed analysis was carried out. A modal analysis was derived by point counting: this involves noting the minerals present in a thin section at a number of points determined by the position of a randomly placed grid. About 1700 points were counted for each section, enough to give statistically accurate results[10]. The results were compared with specimens from known Roman and other quarries, especially from the Wall area. One quarry, however, would often provide such a variety of rock that it was impossible to say for sure, simply on thin section analysis, just which Yoredale sandstone or which St. Bees Sandstone quarry a specimen had come from. Generally, using published petrographic rock descriptions and carrying out fieldwork in selected areas, each specimen was assigned as closely as possible to a stratigraphic horizon and the provenance determined.

4.   THE RESULTS OF THE SURVEY

Southern Area:

Only one milestone (1) survives from the territory of the Cantiaci. It was found near Watling Street west of Rochester and is an undressed sarsen, possibly pillaged by the Romans from a megalithic site.

In the territory of the Regni a stone of Constantine was discovered among villa remains at West Worthing (2). It was probably originally erected on the road running east from Chichester. The stone is from the top of the Lower Greensand in the Washington-Hassocks area. Rivet notes that "in the east of the Regnenses' territory the large cemetery at Hassocks should indicate a centre of some kind ..."[11] and it may be that the stone was quarried there.

Five milestones are extant in the territory of the Belgae. Of these, four (5,9,13 and 14) are Chilmark Stone from the Vale of Wardour, while the fifth (8) is of Bembridge Limestone from the Isle of Wight. Jurassic oolites from the west of the territory and sarsens from the chalk uplands are also suitable for inscribing, as might be some of the Lower Chalk horizons. Hence the use of Chilmark Stone for milestones at Rockbourne, Clanville and Bitterne suggests some degree of control from Venta Belgarum, at least until towards the end of the third century. Certainly we may assume that the Chilmark quarries and Rockbourne were in the civitas of the Belgae for it is unlikely that stone would have been acquired from without when there was ample supply within the territory[12].

In the territory of the Atrebates, the Impstone (4) from west of Silchester and the uninscribed stone from Bannisters to the east (3), are of Bath Stone from the Great Oolite Series a few miles north-east or south-east of Bath. The quarries for this stone, which was also used in buildings at Silchester[13], would have lain within Belgic territory but there is no good stone for inscribing within the Atrebatic area, although stone from some of the Lower Chalk horizons could perhaps have been inscribed.

As expected, the milestone from Venn Bridge on the Fosse Way (16) is of Ham Hill Stone, quarried by the Romans less than two miles from its find spot. Lindinis (probably Ilchester) seems to have reached the status of a second civitas within the area of the Durotriges[14]. The quarry is only about four miles from the town and the milestone may have originally been intended as a column in one of the town buildings.

South-West:

Among the Dumnonii milestones present archaeological as well as geological problems. A probable trunk route through Cornwall is shown on the O. S. Map of Roman Britain (3rd edn., 1956), although it is not recognised by Margary[15]. Although the tin ore was probably removed by sea, the Gordian milestone from Gwennap, near Redruth (19), is local in origin and may well date the reconstruction of this road at the beginning of the development of Roman tin-mining. The Breage and St. Hilary milestones (20 and 21) are also local in origin and probably once stood on relatively minor roads connecting the tin-mining areas around Camborne and at Penwith with a probable port at Porthleven[16]. The milestones from near Tintagel (17 and 18) present a greater problem. The Tintagel example is of local slate, but the Trethevy stone is of granite from Bodmin Moor, at least seven miles south-south-east. It would be odd if the Romans had carried the Tintagel stone from the slate area to the trunk route across the granite uplands of the Bodmin Moor, but it seems even more improbable that a subsequent people would have removed the milestones so far merely to re-use them as gateposts. There must surely have been a coast road by Tintagel and a mile or so inland the ground is suitable, forming an undulating escarpment for several miles. But quite where the road began and where it could have been heading are by no means obvious.

Central Area:

   The milestones from Durobrivae (Water Newton) (24 and
25), those found at Girton, Cambridge (22 and 23), and a
stone of unknown find spot (26) are all of Barnack Freestone which, with the better known Barnack Rag, is exposed
at several places along the Lincolnshire Limestone outcrop.
(See Fig. 2). None is more than 12 miles from Durobrivae
and the Hills and Holes at Barnack are only six miles distant. Closer at hand to Cambridge, Clunch, the name given
to the local harder parts of the Lower Chalk, was quarried
in Roman times for building purposes and, though by no means
as good as Barnack Freestone, it would have taken an
inscription and weathered reasonably well[17]. If we accept
Stevens' argument[18] that Durobrivae was granted the status
of a civitas capital within Catuvellaunian territory it may
be that Cambridge (Durolipons) was included within its area.

   The two milestones from Lincoln are both oolites from
the Lincolnshire Limestone, but apparently from different
quarries. The stone from Sibthorpe St. or High St. (28)
can be matched locally, but the Bailgate stone (29) is
likely to have originated in the Ancaster area, 15 or 20
miles to the south. If this latter stone was quarried within the area of the colonia, as we would expect, we would be
justified in extending the territorium to a radius of nearly
twenty miles and perhaps including Ancaster as well as
Segelocum which is mentioned in the inscription. The milestone found at Ancaster (27) is of local freestone.

   Two milestones found on the Fosse Way near Leicester
were both quarried from the Upper Keuper Sandstone at Dane
Hills, just south-west of the town. This formation was
quarried here for a large Hadrianic building at Ratae[19].
Indeed, the Hadrianic stone from Thurmaston (32) may have
been originally intended for a column of the building
because it has a circular depression in the top perhaps
intended for a mortice, but it is doubtful if it ever
found employment as such. The Six Hills stone (33) from
11 miles north-east of Ratae was probably inscribed in
the same civitas workshop. Unfortunately the inscription
has been almost completely removed and cannot be dated.

   Other milestones from the same area have different
origins. The fragment from Wall (34) is of local Keuper
Sandstone and is similar to the local building stone used
at Letocetum. It seems to be inscribed to Claudius II,
268-70, and by this date a local origin is not unexpected.
The milestone of Constantine from Cave's Inn (Tripontium)
(31) is from the Northampton Sand outcrop. It must have
been quarried several miles to the south or east of Cave's
Inn and probable came from the Roman iron-working centre at
Duston, 18 miles south-east. In this case Tripontium may
have lain within the area of the Catuvellauni and not the
Coritani - unless Tripontium was a new civitas capital as
recent evidence might suggest[20].

Unfortunately the Buxton milestone (35) cannot be dated, though the use of three ligatures in so short a text suggests third century.   It seems to be local in origin; at least it is improbable that it was quarried in the vicinity of the fort at Brough on Noe, whence the distance was measured.

Three milestones from the lands of the Dobunni have different origins. The oldest is the stone of Numerian from Kenchester (Magnis) (40). If the end of this inscription is correctly interpreted as r(es)p(ublica) c(iuitatis) D(obunnorum), then it was set up under the authority of Corinium. The stone, however, is of local Old Red Sandstone from nearby Dinmore or Garnons Hills. Two stones of Constantine are both Jurassic limestones. The Alcester milestone (38) is from the Pea Grit Series which crops out along the Cotswold Edge in the Stanway-Saintbury area. It is likely to have originated from near Saintbury where the road south from Alcester to Bourton on the Water climbs the scarp. The stone from Kempsey was probably quarried at Bredon Hill and may have been transported along an extension of the trackway north from Cirencester to Hales Abbey.

South Wales:

In South Wales five milestones from the coast road fall into two rock groups; three (41, 42 and 44) are coarse grained quartzite (Millstone Grit) of Namurian age, and two (43 and 45) seem to be sandstone from the Upper Coal Measures. The actual quarry has not been found for either type but though both could have come from some distance, it is more likely that they were quarried locally[21]. The Brecon milestone is of local sandstone, probably from the Senni Beds of the Old Red Sandstone.

North Wales:

As would be expected from the preliminary survey, the Hadrianic milestone from Rhiwiau-uchaf (51 - "from Kanovium, 8 miles") and the one inscribed to Severus, Caracalla and Geta (52) found close by, are of Millstone Grit. The coarse sandstone with quartzite pebbles cannot be matched by any outcrop closer than the Cefn-y-Fedw Sandstone which crops out northwards from Llangollen, through Minera and by Mold to Holywell. (See Fig. 3). The pebble size indicates that the stone must have come from the southern part of the outcrop, probably within a few miles of Llangollen. The earlier stone is a good cylinder which has probably been turned; both have neat inscriptions. There is equally good, if not better, rock closer to Kanovium than the Cefn-y-Fedw Sandstone, rock which would take an equally neat inscription and, being without the pebbles, would look better. So although the Hadrianic stone records the distance from Kanovium, it would seem that, as might be expected from Mommsen's analysis, there was some central control over road maintenance and milestone erection throughout the North Wales peninsula by the legionary fortress at Chester. The milestones may have

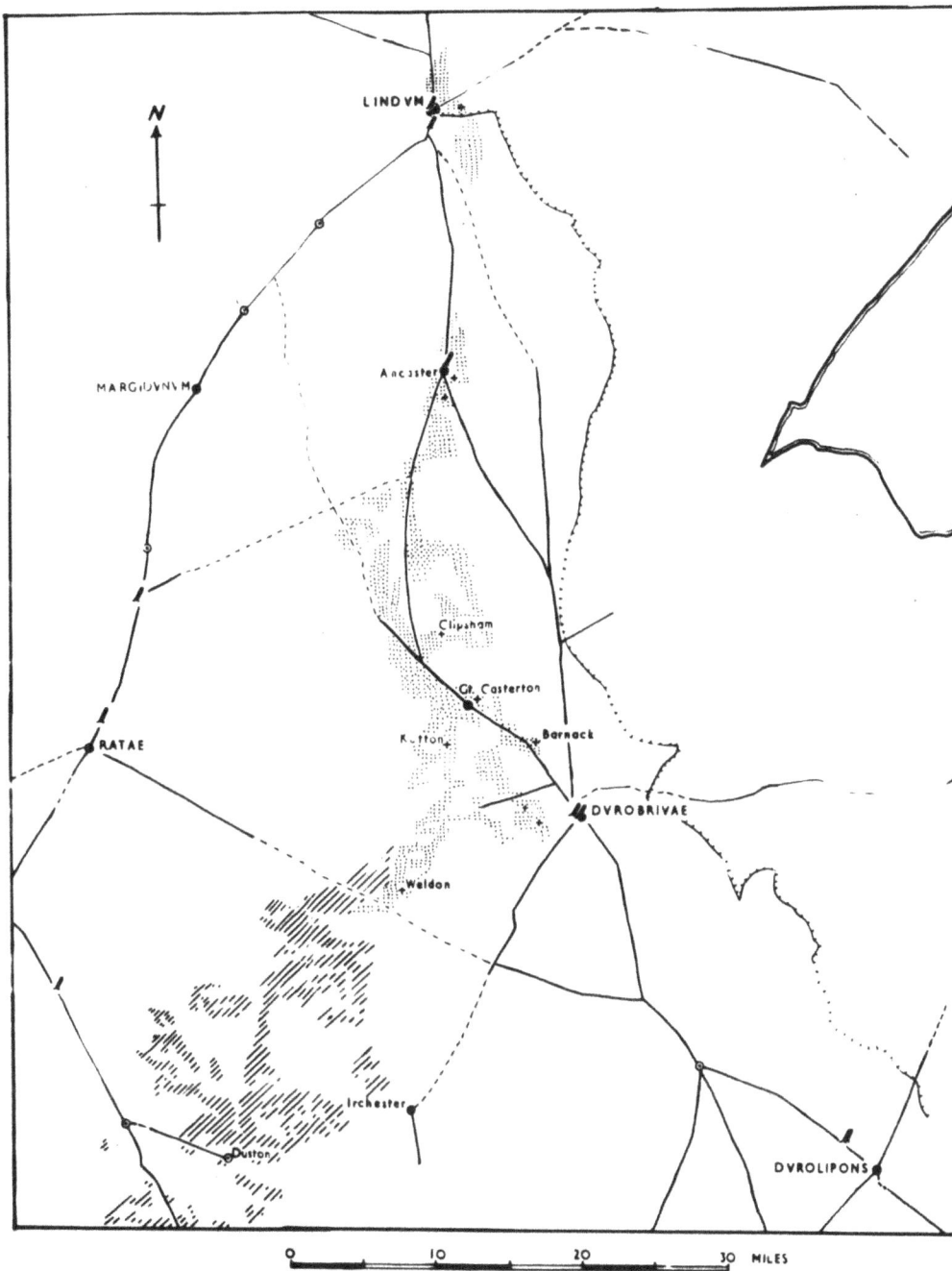

N

LINDVM

MARGIDVNVM

Ancaster

Clipsham

Gt. Casterton

Rutton

Barnack

DVROBRIVAE

RATAE

Weldon

Irchester

Duston

DVROLIPONS

0   10   20   30   MILES

LEGEND

MILESTONES../
TOWNS ....●
LARGER SETTLEMENTS............⊚
ROMAN CANAL AND
    NAVIGABLE RIVER ....

LINCOLNSHIRE LIMESTONE
(better stone areas) ....
MAIN QUARRY SITES.....................+
NORTHAMPTON SAND...............⧄

FIG. 2   Lincolnshire limestone and Northampton
          sand outcrops in relation to the Roman
          geography and milestones.

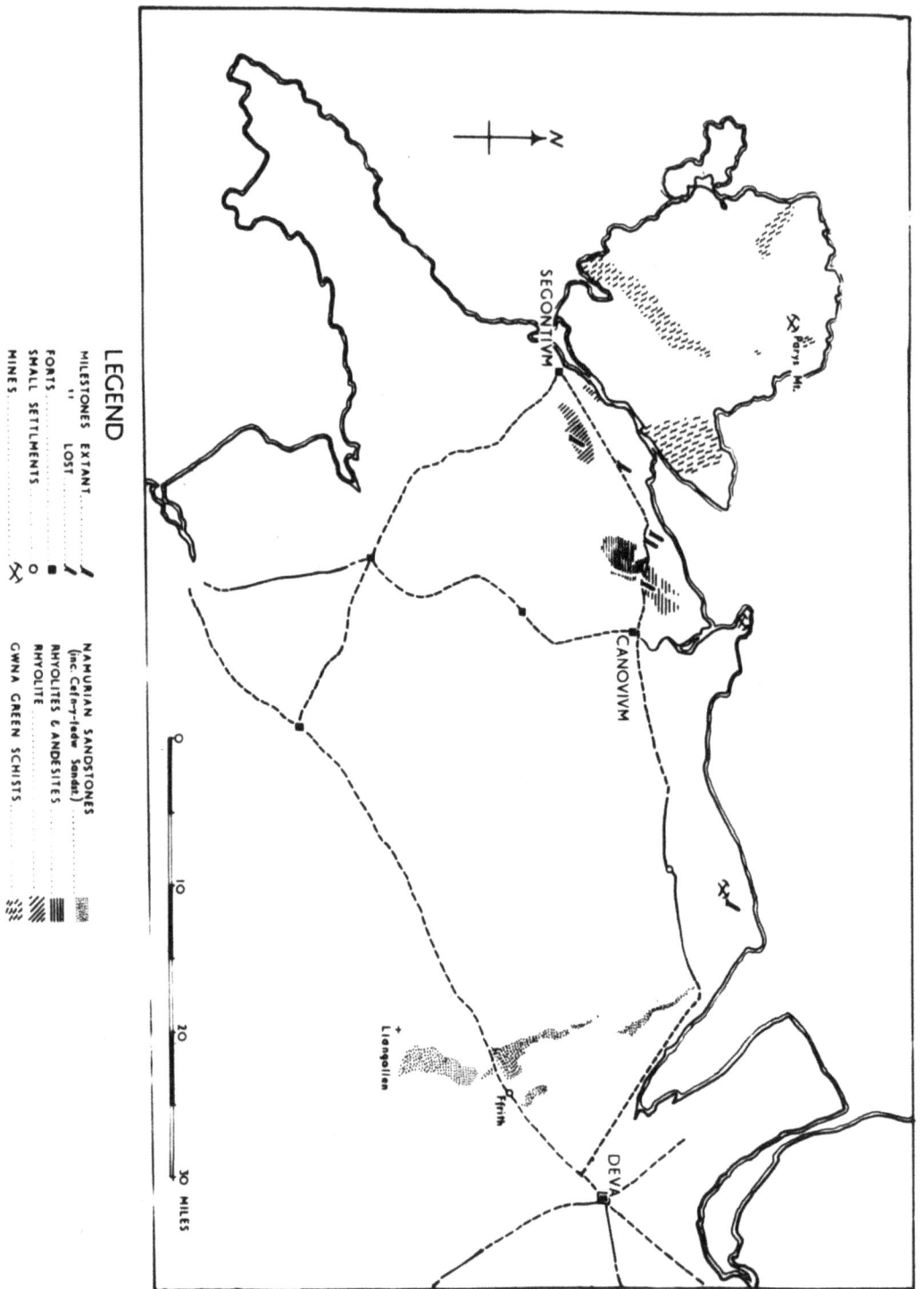

LEGEND

MILESTONES EXTANT. /
           " LOST /

FORTS ■ □

SMALL SETTLEMENTS ○

MINES ⚒

NAMURIAN SANDSTONES (inc. Cefn-y-fedw Sandst.)

RHYOLITES & ANDESITES

RHYOLITE

GWNA GREEN SCHISTS

SEGONTIVM

CANOVIVM

DEVA

Llangollen

Ffrith

Parys Mt.

0   10   20   30 MILES

FIG. 3 North Wales: the milestones and relevant rock outcrops.

been completed at the quarry and transported along some unknown road to the North Wales coast road. More probably the crude milestones would have been transported to Chester, perhaps along the River Dee, where they were inscribed before being taken by road or sea to Caerhun.

Of the other milestones from North Wales, that of Severus Alexander from Gwaenysgor (49) is a medium-grained dolerite. There is no known local source of dolerite which could have supplied this stone and it must have come from further west in the Snowdon area or perhaps from Anglesey. It is not a good stone for inscribing and yet the lettering is quite neat, suggesting that it was produced by competent masons. It is possible that it found its way east to Chester in a boat bringing copper ore from Parys Mountain and was utilised simply because of its convenient shape; but it is more likely that it was inscribed at Segontium and brought east during road reconstruction.

The use of this poorer quality stone may foreshadow a change in the organization of road repair in North Wales. During the early years of the third century the legionary bases at Caerleon and Chester were renovated and repaired as part of a plan to strengthen the Welsh frontier. At the same time certain outlying forts, especially Caernarvon (Segontium) and Caersws, seem to have grown in importance relative to the civil administration of the local tribal areas[22] and road maintenance could well have become one of their responsibilities.

The remaining milestones of North Wales are all local in origin. The stone from Llys Dinorwic (55) is a local rhyolite. The stone of Postumus from Aber (53) is a green schist from Anglesey. It may have been quarried as a milestone but more probably it found its way across the Menai Straits as ballast in a boat and was utilised because of its shape. The stone of Constantine from Bwlch-y-Ddeufan (50) is a very hard, fine-grained igneous rock, so hard that the mason must have had a very difficult job in incising even the very shallow inscription that it carries. The stone is a natural fragment and must have been used only because it was right on the spot: there is better stone for inscribing but a mile or two distant.

Northern Area:

The remaining Hadrianic milestone (61), from Caton near Lancaster, is more difficult than the other two to place, both as regards its origin and the authority responsible for its erection. In this area a military authority based on a fort, rather than a civil authority based on a town, would have had responsibility for road maintenance. The stone is of Millstone Grit. Its origin could be almost anywhere in the Lancaster Fells and Forest of Bowland area; perhaps the stone was from very close at hand, but it could have come from as far away as Longridge, four or five miles north of Ribchester, where a reasonable match has been found. The stone is believed to measure the distance from Lancaster, though the reading is doubtful.

Stones of Philip and Decius (59 and 60) were found close to each other three miles south of Lancaster. The former is of Bunter Sandstone from some miles further south, towards the River Ribble. The latter is probably of local Millstone Grit and may well have been a loose weathered rock which was crudely and carelessly inscribed after Decius's ascendancy. The fragmentary stone from Ribchester, inscribed to Decius (57), was probably quarried a mile or two north on Longridge Fell, though it is of different material from the Hadrianic milestone. The other stones from the Lancaster area, at Overtown Farm (62) near Burrow in Lonsdale and at Middleton (63), are Carboniferous sandstones with probably local origins from the Yoredale facies.

Carlisle possessed an early fort and later seems to have become the centre of a new civitas in what was formerly Brigantian territory[23]. It is reasonable to expect, though, that some other fort was made responsible for maintaining communications in a distant part of the territory, as seems to be the case with the Ingliston (108) and Buxton (35) milestones. In the southern part of the new civitas it is not clear whether this responsibility fell on Burrow in Lonsdale (Galacum), Lancaster or Ribchester (Bremetennacum).

The milestones from the area of Isurium Brigantum are all dated to the reign of Decius or later. The stones of Decius from Duel Cross (66) and Aldborough (67 and 68) are of Keuper Sandstone, probably from the quarries at Aldborough which also furnished building stone for the Roman town. Milestones from Stainmore (74, 75 and 76) and Greta Bridge (73) are of Carboniferous sandstone, probably from the Namurian, or perhaps from the Lower Coal Measures (Westphalian), and are also local in origin. The town or fort to which the Stainmore stones belong is not certain but geographically they relate to Isurium or Eboracum rather than to Corstopitum or Luguvalium. The Piercebridge milestone (69) is a quartzite rock, most likely local, though no reasonable match has been found. Further south, the Castleford milestones, although each is of a different rock, are fairly local in origin. That from Beancroft Road (64) is of a quartzite similar to the Piercebridge milestone, and the stone from Carlton Street is cut from a softer, felspathic sandstone, probably from the Upper Namurian around Huddersfield and Halifax.

In the vicinity of Carlisle the milestones fall geologically into three groups. Milestones from Brougham (80 and 81) and Temple Sowerby (79) are of local Penrith Sandstone, probably quarried in the hills near Brougham. Both the Scalesceugh stone (84) and the stone from the Military Way at Old Wall (87) are composed of St. Bees Sandstone. They are both well matched by the St. Bees Sandstone from the Roman quarry at Wetheral[24], but they

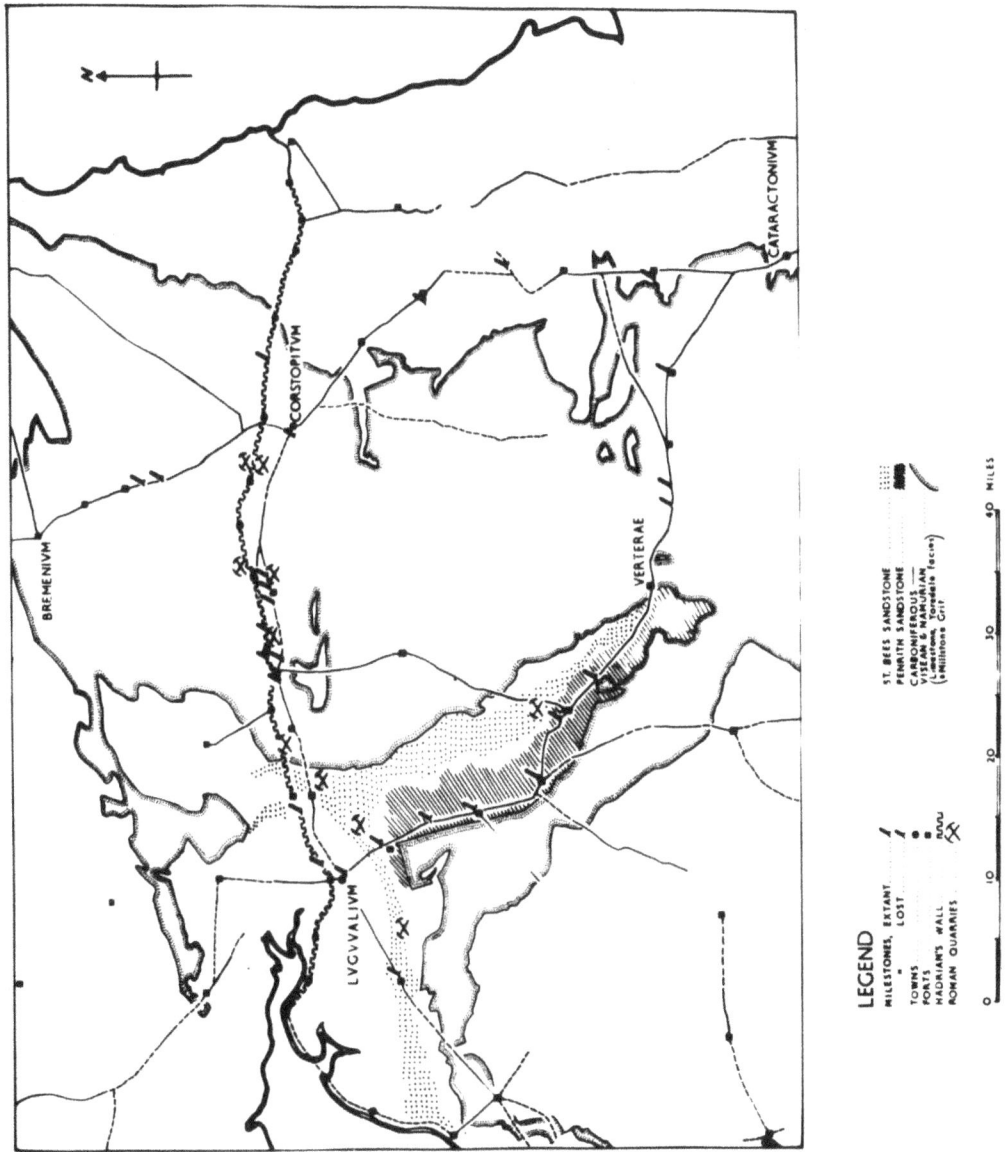

LEGEND

MILESTONES, EXTANT	↗	ST. BEES SANDSTONE
" LOST	╱	PENRITH SANDSTONE
TOWNS	●	CARBONIFEROUS
FORTS	■	VISEAN & NAMURIAN
HADRIAN'S WALL		(Limestone, Teesdale facies)
ROMAN QUARRIES	✗	(Millstone Grit)

0   10   20   30   40 MILES

FIG. 4 Hadrian's Wall area: milestones,
        relevant rock outcrops, and Roman quarries.

could have come from other places within the outcrop.
(See Fig. 4). The Aurelian milestone from the Stanegate
at Fell End, near Carvoran (89), is also of the same St.
Bees Sandstone. The remaining milestones, from Hangingshaw, near Appleby (78) and at Harraby Bridge (85), are
of similar quartzite, seemingly of Carboniferous age.
The origin of these milestones is by no means certain
and the significance of two similar stones being found so
far apart is far from clear.

With the exception of the Old Wall and Fell End milestones, all the stones from the Wall area are of local
Carboniferous sandstone from the Yoredale facies. The
various sandstones have been extensively quarried in the
past, not least by the Romans for building Hadrian's Wall.
Because each sandstone is very much like another and
because each horizon may vary to some extent in composition
and in a regular fashion from coarse to fine grain, it is
not usually possible to be certain from which horizon, and
still less from which quarry, a specimen has come.

There are two milestones inscribed to Severus Alexander in the area, one from the Stanegate at Crindledykes
(92) and the other from the Military Way at Cawfields
(102). Though the Crindledykes stone has weathered to a
deeper yellow than the other, as far as grain size and
composition are concerned they could well have come from the
same quarry. If we note also that the inscriptions have
very nearly the same style of lettering and that both
stones are of nearly the same dimensions and both record
distances from Corstopitum, it seems safe to conclude that
they came from the same quarry and both were inscribed
by the same workshop.

Of the other stones, it is possible that the milestone
of Numerian from Cawfields (103) and perhaps also that of
Maximinus from Crindledykes (94) could be composed of the
same stone as the Severus milestones. The worn milestone
of Constantine from Crindledykes (96) and that of Maximinus
from Corbridge (100) may share the same origin, as also may
the Crindledykes stones of Probus (93) and Constantine as
Caesar (95). It is not possible to say more than this.
Samples of stone from the Roman quarries at Haltwhistle Burn
and Bruton both provide possible matches for the milestones
of Severus and perhaps also for the other milestones, but
other exposures would also match.

The Antonine milestone from Ingliston (108) is in-
complete and the distance has been lost from the inscription,
but the caput viae, Trimontium, remains. It is likely that
the stone originated from the Upper Old Red Sandstone of
Lauderdale or nearby, in which case the milestone must have
been quarried and erected under the military authority at
Trimontium (Newstead). Richmond[25] has noted that Newstead
was clearly the dominant fort in this part of the Scottish
Lowlands and that it was reconstructed in the early part of
the Antonine period to take a detachment of the Twentieth

14

Legion based at Chester. It is not surprising, then, to
find the fort taking responsibility for road-works, at
least in the eastern Lowlands, and perhaps also for the
Antonine Wall area. There are no other milestones from
this area, but the lower portion of a similar stone, also
measuring the distance from a fort, was found at Buxton
(35 and p.8).

5.   SUMMARY

Decentralisation of authority. There is only one early
milestone from the civil area - the Thurmaston stone
quarried and found close to Ratae (p. 7) - and here there
are no later milestones for comparison. In the military
area, if the Gwaenysgor milestone in North Wales was in
fact produced at Segontium and carried east (p.11) to be
placed on the coast road then that fort may, by the middle
of the third century, have become responsible for the
maintenance of roads in North Wales as far east as
Prestatyn. We may be seeing, though, no more than the use,
for purely economic reasons, of a convenient and inferior
quality stone. But if the evidence of the Ingliston
milestone (p. 14) is correctly interpreted it might add
weight to the idea of late second or early third century
decentralisation. Though we are not helped to distinguish
between the different types of road that Mommsen noted in
North Africa (p. 1), this evidence for decentralisation
is in line with his observations.

     In the period from the early third century onwards we
still find that, at times, care was taken to produce good
milestones. In the north, the Stainmore milestones, though
of local material, were carefully prepared and must have
been produced by competent masons. Further south, in the
territory of the Belgae, good stone was still being quarried
at, and carried some distance from, the quarries at Chilmark.
But by this time, in most areas, local and often poor stone
was being incised more or less on the spot.

Civitas boundaries. At one or two points the milestones can
be used to delimit the civitas boundaries (see Fig. 5). In
the south the Chilmark quarries and Rockbourne villa seem to
be included in Belgic territory (p. 6). Further north,
Cave's Inn is likely to have been in the territory of the
Catuvellauni, unless it did itself become a civitas capital
(p. 7). If Durobrivae (Water Newton) was elevated to the
status of civitas capital, it is possible that Durolipons
was included in its territory (p. 7). Also there is reason
to suppose that the area of the colonia at Lincoln should be
extended to nearly twenty miles radius (p. 7). Only two
milestones, the Impstone and Bannisters milestone, seem
definitely to have come from outside the civitas territory
to which they belonged (p. 6), and the lack of other
suitable stone explains this.

15

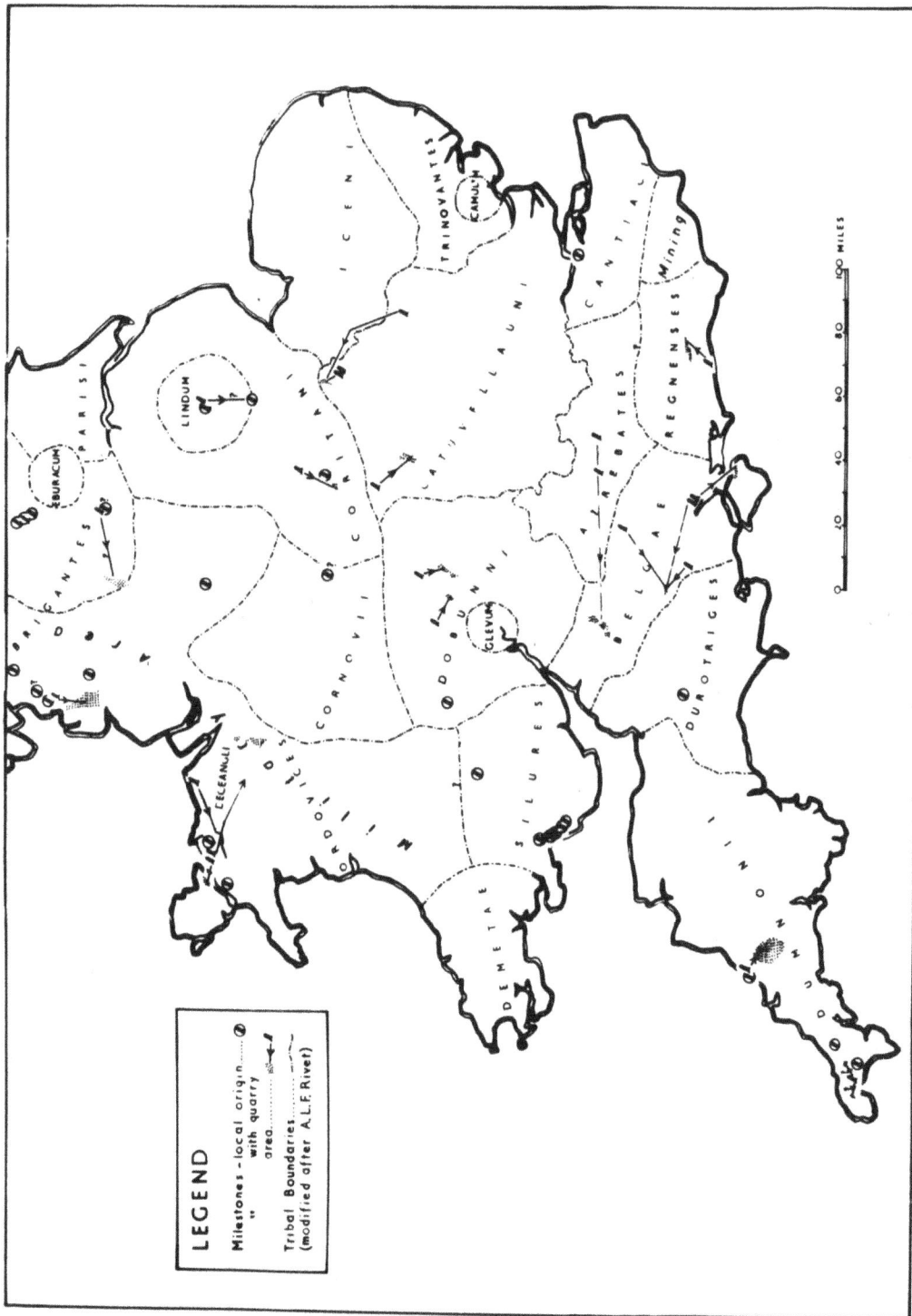

FIG. 5   Southern Britain: the milestones, their
         origins and civitas boundaries.

APPENDIX

CATALOGUE OF MILESTONES

The milestones are listed below more or less in the
order given in R.I.B., i, with appropriate modifications
to suit the needs of political geography and consistency,
and with additions where there are uninscribed milestones
and milestones found since 1955. Full references are not
given with inscribed stones. Milestones which are now
lost have been included for the sake of completeness.

Where possible each milestone is given a six figure
National Grid reference for its find spot; where
feasible and appropriate an eight figure reference is
given. Where stones have been found on or close by a
Roman road, the Margary road number is given[26].

1 SHORNE

R.I.B. 2219

TQ 695712

FLAVIUS SEVERUS & MAXIMINUS
DAIA: 305-6

Irregular: 117x56x30.5 cm

Found 1907 near The Warren, Shorne, about 3 miles west of
Rochester and 1¼ miles north of Watling Street (road 1c); now in
Maidstone Museum. Lettering reasonably neat, though shallow;
some letters nearly worn away. A modern inscription, THE WARREN/
SHORNE, has been carved over lines 4 and 5.

A well sorted orthoquartzite, hard and flinty on the outside
and weathered orangy-red. An original sarsen which has not been
cut down or trimmed in any way.

Sarsens, or "grey-wethers", often occur in the chalk areas of
the south and in the associated Pleistocene and Recent deposits.
It is most probable that this stone came from near Chatham and more
particularly from between Chatham and Bredhurst where, in the
Pleistocene clay-with-flints, they are often iron stained[27].

It is unusual that this milestone is not dressed in any way,
and it would be a little odd if the Romans had troubled to carry
it eight or ten miles only to leave it undressed. Jessup records
that at Battle Street, a hamlet of Cobham, about half a mile south
of Watling Street and about two miles from The Warren, there is
the site of a megalithic stone circle known as the Druid's Temple
or Warrior's Grave[28]. The stones have, in more recent times,
been removed to form steps or boundary stones in the neighbourhood, and we may reasonably conjecture that at least one of the
stones found later employment as a Roman milestone on Watling
Street.[29]

2 WORTHING

R.I.B. 2220

TQ 133028

CONSTANTINE I: 307-37

Rectangular: 96.5x53x20 cm

Found 1901 in remains of a Roman villa at Herschel Lodge,
Worthing, a mile or so south of the road (153) east from
Chichester; now in Worthing Museum. Lettering a little defaced but otherwise moderately neat. Lower portion of a stone
broken into three pieces.

A lithic, glauconitic greywacke, poorly sorted with occasional plagioclase grains. A variety of iron ores are present, mainly
glauconite, occurring as primary detrital and secondary grains.
Stone a grey-yellow, but with orangy-red and greyish streaks.

A greensand and probably from the top of the Lower Greensand,
the Folkestone Beds, which provides a good match a few miles north
of Worthing in the Washington-Hassocks area[30].

3 BANNISTERS

V.C.H. Berks., i (1906), 207[31]

SU 782639

UNINSCRIBED

Rough Cylinder

Found, reputedly in 1841, in a field called Six Acres on Webbs
Farm, Finchampstead, Berks.; now at Bannisters, Finchampstead.
Originally on road (4a) from London to Silchester. Badly weathered
but may once have had an inscription.

A light buff, shelly oolite; some shell fragments quite
large. There is a partly obscured terebratulid of ? Bathonian
age. Ooliths are regular, nearly 1 mm in size. Fragments,
ooliths and micrite matrix are partly recrystallised to fine
grained sparry calcite. Some siderite present.

The rock is coarser and more shelly than the Impstone (4) but is otherwise similar, e.g. in amount of micrite matrix and extent of re-crystallisation. It is Bath Stone from the Bath area where the Great Oolite forms good, pale buff, well cemented freestones and shelly oolites[32]. It has been frequently quarried in the past and was used in some of the buildings at Calleva[33]. There is no local stone which is suitable for inscribing as a milestone, though there are outcrops of the Upper Greensand west from Burbage to Devizes which may have provided suitable material.

4 IMPSTONE   R.I.B. 2221

   SU 60966247   UNDATED

Roughly rectangular: 30.5 cm above ground x46x28 cm

Recorded by Stukely in 1776 and by the Ordnance Survey in 1838, 1¼ miles due west of the west gate of Silchester, on the north side of the modern road; still at this spot. Re-used as a parish and county boundary stone and with a bench mark on it. Wright suggests that "it is quite possible that it was once a Roman milestone. If so it must have been moved ½ mile from the line of the Roman road which runs west-north-westwards towards Thatcham" (Road 41a). However the road on which it now stands is straight for 3½ miles before turning north-west in a straight alignment towards the Thatcham road. As this road also forms the parish and county boundaries, it may also be the line of a Roman road, possibly an early one.

A light buff, shelly oolite. Ooliths, though variable in size, do not usually exceed 1 mm; shell fragments may be 3.5 mm or more. Some recrystallisation of the ooliths, fragments and partly micritic matrix. Traces of siderite present.

Very similar to the Bannisters milestone (3) and also Bath Stone and not Greensand as stated in R.I.B. It is possible that they have both come from the same quarry, though they would be from different horizons.

5 CLANVILLE   R.I.B. 98

   SU 31454898   CARINUS: 282-3

Rectangular: 33x35.5x28 cm

Found 1897 10 yards south of the villa at Clanville, Weyhill, Hants.; now in Winchester Museum. The resemblance of this stone to a milestone has long been recognised but in the past, because of its provenance from a villa 3 miles from the nearest Roman road (43 and 4b), it has usually been considered to be an honorific inscription. Since the discovery of the milestones at Rockbourne, nearly 5 miles from the nearest Roman road, there is better case for recognising the Clanville stone also as a milestone.

Inscription neat but shallow. There is a small gap between the third line of the inscription and the base where the stone has been cut across in re-use. Although there is no trace of a fourth line there could well have been more inscription on the lower portion of the stone. The front corners are slightly rounded and the inscribed face slightly hollowed suggesting that this might be a second inscription.

A pale, buff-grey, fine-grained, sandy limestone with recrystallised, abraded shell fragments and well rounded sand grains. Some green glauconite present, sometimes oxidised to limonite.

It is Chilmark Stone from the Upper Portland Beds of the Vale of Wardour which has been exploited in many quarries around Tisbury, Wardour and Chicksgrove, as well as at Chilmark[34]. There are two main stone horizons which vary a little in composition but are essentially similar. The lower horizon, which has been widely used for building purposes, is further subdivided into the Pinney, Trough and Green Beds. Obviously there will be some variation in stone which is termed Chilmark Stone, as there is variation from bed to bed and from quarry to quarry. This milestone is probably from the Green Bed, though it could be from the Pinney Bed.

6 BITTERNE 1	R.I.B. 2222

GORDIAN III: 238-44

Rectangular: 71 cm high

Found 1804 or 1805 in the town walls of Bitterne (Clausentum); now lost. Inscription apparently mentioned the authority responsible for its erection - the civitas of the Belgae[35].

7 BITTERNE 2	R.I.B. 2223

GALLUS AND VOLUSIAN: 251-3

Roughly rectangular: 84x38 cm

Found 1804 or 1805 in the town walls of Bitterne; now lost. The stone tapered to a gable.

8 BITTERNE 3	R.I.B. 2224
SU 435134	i GORDIAN III: 238-44
	ii TETRICUS: 270-3

Roughly oval: 109x46x28 cm

Found shortly before 1845 in the town walls at Bitterne; now in God's House Tower Museum, Southampton. Apparently inscribed twice, the more complete second inscription now picked out in red. Back and right side partly cut down in re-use and traces of mortar on the back.

Fine grained, light grey or creamy limestone composed entirely of micrite except for fine patches of sparry calcite in or around recrystallised fossil fragments. Stone pitted where fossils have weathered out but traces of small gastropods can still be seen.

Bembridge Limestone from the Oligocene of the Isle of Wight and probably from the Binstead Limestone, a variety quarried at Binstead, especially during the Middle Ages, and now worked out[36]. From Binstead it could easily have been shipped to Clausentum, or even further inland. (It is geologically unlikely that the stone could have come from the small outlier of Bembridge Limestone at Creechbarrow, 2½ miles west of Castle Corfe, Dorset).

9 BITTERNE 4	R.I.B. 2225
SU 435134	TETRICUS: 270-3

Rectangular: 74x35.5x20 cm

Found 1841, or just before, at Bitterne, in the garden of the manor; now in God's House Tower Museum, Southampton. Lettering shallow but neat. Back rough hewn with traces of mortar from secondary use; trimmed down at the edges.

Sandy, glauconitic limestone which appears pale buff-grey. Shell fragments abraded and recrystallised: quartz grains moderately angular. Glauconite often altered to limonite; some siderite.

Chilmark Stone from the Upper Portlands Beds of the Vale of Wardour. Probably from the Green Bed. (See 5, above, for details).

10 BITTERNE 5	R.I.B. 2226

TETRICUS: 270-3

Squared stone: 38x38 cm

Found 1804 or 1805 in the town walls of Bitterne; now lost. Square, neat stone cut down from a milestone.

11 BITTERNE 6   R.I.B. 2227

                                 AURELIAN: 273-5

Squared stone: 46x30.5 cm

     Found 1799 at Bitterne; now lost. Perhaps a milestone.

12 BITTERNE 7   R.I.B. 2228

                                 UNDATED

Cylindrical: 92x46 cm diam.

     Found 1804 or 1805 in the town walls at Bitterne; now lost. The
lower part of a milestone from which the emperor's name had been lost.
The inscription ran "...in the eighteenth year of his tribunician power
restored the roads which had fallen into ruin through age..." Haverfield
(E.E. ix) has suggested that it may perhaps refer to Severus, 210, or
Caracalla, 215, for it does not resemble the inscriptions of the fourth
century emperors who held tribunician power for the eighteenth time.
Birley[37] has observed that by comparison with the stone from Rhiwiau-
uchaf (52) Severus would also have his sons mentioned on the stone, and
so prefers the date of 215. Either date fits adequately the known period
of road building and renovating at the beginning of the third century.

13 ROCKBOURNE 1   J.R.S., lvi (1966), 219

     SU 117170   DECIUS: 249-51

Rectangular: 51x28x20 cm

     Found 1965 in the foundations of room VI of West Park Villa,
Rockbourne, Hants.; now in the site museum. This is nearly 5 miles from
the nearest known Roman road (4c). Lettering shallow but very neat with
the mason's guide lines still visible so that the stone can hardly have
been exposed to the weather. The back is worn smooth where the stone
has been secondarily used as a step and it is likely that it was never
used for its intended purpose, almost certainly because Decius was over-
thrown before it could be erected. Top irregular but original; base
broken.

     Buff, sandy limestone. Quartz grains fairly angular; calcite often
occurs as clastic grains. Some greenish glauconite occasionally altered
to limonite.

     Chilmark Stone, probably from the Pinney Bed of the Upper Portland
Beds in the Vale of Wardour. (For details see 5).

14 ROCKBOURNE 2   J.R.S., lii (1962), 195[38]

     SU 117170   TETRICUS: 270-3

Rectangular: 94x38x23 cm

     Found 1961 in the footings of a flint wall on the north side of
room XIV at the west end of the Roman villa at West Park, Rockbourne,
Hants.; now in the site museum. Inscription to Tetricus or both
Tetrici; neat but shallow. Broken at top; damaged at back and a little
at sides.

     A buff, slightly glauconitic, sandy limestone. Calcite occurs as
abraded shell fragments, ooliths and micrite matrix, all partly recrystallised. Quartz grains fairly angular. Glauconite sometimes
altered to limonite.

     Chilmark Stone from the Upper Portland Beds of the Vale of Wardour,
probably from the Green Bed. (See 5 for details).

15 STINSFORD

Pope, *Old stone crosses of Dorset*, (1906), 177[39]

SY 70899130   UNINSCRIBED

Rough cylinder: 180 cm above ground x 38 cm diam.

First noted by Pope at Stinsford, Dorset, on line of Roman road (4e) from Dorchester (Durnovaria) to Badbury (Vindocladia); moved at least twice during road reconstruction but probably not more than a few feet from its original spot. It tapers slightly towards the top. It has signs of wear from a chain which has led to the belief that it was the stone known locally as the "bull-baiting stone". It may have been so used, or as a rubbing stone, but whether or not such a function was primary or only secondary, it is not possible to say. Its shape and location are both consistent with its primary function as a Roman milestone.

This stone has not been sampled but it is a grey or grey-buff, shelly, slightly oolitic limestone, probably of fairly local origin.

16 VENN BRIDGE   R.I.B. 2229

ST 478188   FLAVIUS SEVERUS: 305-6

Cylindrical: 123x29 cm diam.

Found 1930 on south bank of stream at Venn Bridge, Stoke sub Hamdon, 5 miles south-west of Ilchester on the Fosse Way (5a); now in Taunton Museum. Lettering rough, fairly deeply cut. It has a moulded base, 33 cm diam. Apparently primarily intended for use as a colonnade or statue base. It has a natural vertical fracture from front to back and it is quite possible that this developed during preparation of the stone as a column, causing it to be used as a milestone instead. Back still smooth but inscribed face and sides chiselled down and rough: present inscription almost certainly a second one.

A buff-orange sandy limestone: sparsely oolitic but containing many cominuted and abraded shell fragments arranged roughly along the bedding.

I was not able to sample the stone for microscopic analysis but it seems to be Ham Hill Stone from the Yeovil Sand, Upper Lias. This stone was quarried in Roman times at Ham Hill, little more than a mile from the find spot[40].

17 TRETHEVY   R.I.B. 2230

SX 076891   GALLUS & VOLUSIAN: 251-3

Rectangular pillar: 135x35.5x28 cm

Found 1919 re-used as a gatepost on the site of the reputed monastery of St. Piran, 1½ miles east of Tintagel; now resting in a paving, leaning against a wall of the house, "St. Pirans", near where it was found. Inscription badly weathered and almost unreadable.

A moderately coarse, biotite granite, with feldspar phenocrysts up to 2 cm long. Quartz fairly fresh but feldspars, mainly orthoclase, often sericitised. Mica is mainly biotite but there is some muscovite. Some tourmaline present.

This is one of the South-West granites which are all essentially similar in composition[41]. The relatively small size of the feldspar phenocrysts favours the Bodmin Moor or Carnmenellis bodies while the relative proportions of the different feldspars indicates the Bodmin Moor or Dartmoor masses[42]. Since the Bodmin Moor granite crops out at its nearest point only 7 miles or so to the south-south-east of the find spot, it is reasonable to conclude that it has come from that mass.

18 TINTAGEL   R.I.B. 2231

SX 051884   LICINIUS: 308-24

Rectangular pillar: 150x30.5x20 cm

Found 1889 built into a stile at the east end of the churchyard at Tintagel Church; now cemented against the wall of the south transept in the church. Inscription reasonably neat but shallow and weathered. The top is now worn to a point where it has been used for sharpening tools in later times.

A dark, blue-grey siltstone with a slight tendency to be slaty. Quartz grains have their long axes oriented sub-parallel; muscovite flakes also tend to lie parallel. Quartz grains have moderate rounding and sphericity.

This is almost certainly from the local Upper Devonian beds which in this area are normally blue-grey in colour and vary from very low grade phyllites, essentially siltstones and mudstones, through to low grade schists[43].

19 GWENNAP   R.I.B. 2234

   SW 720419   GORDIAN III: 238-44

Rectangular pillar: 110x35.5x20 cm

Found 1940 while ploughing at Mynheer Farm, in a field 300 yards east of Gwennap Pit and 1¾ miles due east of Redruth; now re-erected in the garden of Mynheer Farm. Inscription reasonably neat but badly weathered. Broken at sides and back.

A coarse, brown-stained granite with slightly pink feldspar pheno-crysts 2-3 cm in length. Feldspar is mainly orthoclase. Micas are present and there is a little tourmaline.

This rock is typical of the outer portions of the South-West granites; there is nothing sufficiently distinctive to indicate from which it has originated. It is quite possible, therefore, that it was quarried very locally from the small Carn Marth outcrop, or just two miles or so to the south-west from the main Carnmenellis body[41, 42].

20 BREAGE   R.I.B. 2232

   SW 618285   POSTUMUS: 258-68

Rectangular pillar: 170x41x25 cm

First noted 1920 re-used as a gatepost about 100 yards from Breage Church, near Helston; now resting against a wall in the church. Lower portion of inscription removed; lettering crude and badly worn. It has rounded edges suggesting that the present inscription may be secondary.

A medium to coarse grained biotite granite with feldspar phenocrysts up to 3 cm in length. Feldspar is mainly potash feldspar. Green and brown varieties of biotite are present and are more abundant than muscovite. Tourmaline is a common accessory mineral.

The origin of this stone cannot be stated with certainty for it could have come from any of the South-West granites. Close at hand to the find spot are the masses of the Godolphin and Tregonning Hills just to the west and, a mile or two to the north-east, the large Carnmenellis mass. The relatively small size of the feldspar phenocrysts, the quantity of tourmaline and the other modal proportions are all consistent with the stone having come from either of these two masses[41, 42].

21 ST. HILARY   R.I.B. 2233

   SW 550313   CONSTANTINE I: 306-7

Rectangular: 127 cm above floor x 58x30.5 cm

Found 1854 built into the fourteenth century foundations of the chancel of St. Hilary's Church, Cornwall; now re-erected in the south aisle with its base cemented into the floor concealing the presumably uninscribed portion of the face. Lettering moderately neat but weathered.

A fine-medium grained granite. Some feldspar phenocrysts about 2 cm in length. Quartz fairly fresh, occasionally with tiny inclusions of topaz and tourmaline. Feldspars often sericitised. Micas are muscovite and biotite.

This stone may have come from one of the finer granites which occur within the main masses. The nearest localities are at Crowan Beacon, 6 miles north-east on the Carnmenellis mass, or at Castle an Dinas, by Chysauster, about 3 miles north-west in the Land's End granite. There are also finer grained parts in the Godolphin-Tregonning Hill mass, 3 or 4 miles to the east. It is not possible to say which of these three supplied this stone but all are fairly local[41], [42].

22 GIRTON 1   R.I.B. 2237

   TL 409620   CONSTANTINE I: 305-6

Roughly rectangular: 79x35.5 cm high

   Found 1812 with 23 at Girton, nearly 3 miles north-west of Cambridge, by the road (24) to Godmanchester; now in the Museum of Archaeology and Ethnology, Cambridge. The stone has a black encrustation; lettering badly weathered. It seems to have been inscribed twice with the same inscription[44]. Probably originally cylindrical, now roughly rectangular with rounded edges, especially on the right side where some of the lettering extends from the face to the side. It is likely, therefore, to have been an earlier stone re-cut during Constantine's caesarship.

   A light buff, fairly fossiliferous, oolitic limestone. Ooliths generally fairly well rounded, about 0.8 mm diam. Abundant recrystallised shell fragments of sparry calcite up to 5 mm long. Some micrite matrix and traces of siderite and limonite. A medium grained sparry calcite cement.

   A typical Jurassic oolite from the Inferior or Great Oolite. The nearest and most likely formation from which it has come is the Lincoln-shire Limestone, Inferior Oolite, which crops out from Kettering, north-east to the Stamford area, and thence north through Lincoln, to die out towards the Humber[45]. It varies both vertically and laterally, from even-grained oolitic freestones, as at Ketton, to very shelly, hard, raggy bands, as at Barnack, with many grades of shelly oolites in between. At many places along its outcrop it has been quarried for building purposes, the freestones usually providing the best stone though the rag horizons have not been neglected. Indeed, at the well known Barnack quarries, near Stamford, it was the rag which formed the main output. The rag was quarried as early as Roman times and although the freestone, here a shelly oolite, was still available when the rag ran out in the fifteenth century, the quarries were closed[46]. A specimen of shelly oolite from the Hills and Holes at Barnack has a few grains of silt-grade quartz but otherwise provides a good match for this stone from Girton. It is to be expected that the freestone rather than the rag was used for the milestone since it would take a better inscription.

   Shelly oolites and freestones occur both above and below the rag at Barnack and can be traced with the rag south-west to Weldon where it has been quarried at least since the Middle Ages, and north to the Clipsham area[47]. Specimens from these places, though similar, do not provide such a good match as the Barnack Stone but they cannot be entirely ruled out. However, the proximity of the Car Dyke to Barnack and a good means of transport perhaps favour this origin.

23 GIRTON 2   R.I.B. 2236

   TL 409620   UNDATED

Roughly rectangular: 84x35.5x18 cm

   Found 1812 at Girton, Cambridge, with 22; now in the Museum of Archaeology and Ethnology, Cambridge. As with 22, the inscription is a palimpsest but, as the stone has lost its top portion, neither records an emperor's name. Lettering badly weathered; the stone has a black encrustation. Probably originally cylindrical, now roughly rectangular with rounded edges.

   A light buff, shelly oolite. Ooliths small and equal in size, about

0.8mm diam. Shell fragments largely recrystallised, up to 5 mm in length. Some micrite matrix and a high proportion of sparry calcite cement.

Same stone as 22 and well matched by the Barnack Stone from Barnack, south-east of Stamford.

24 DUROBRIVAE 1   R.I.B. 2238

   TL 11979718   VICTORINUS: 268-70

Rectangular: 86x46 cm at top x38 cm

Found 1913 on the south-west side of the Roman road (2b) in a field called The Castles, about 110 yards from the north-west angle of the Roman town; now in Peterborough Museum. Approximately rectangular tapering slightly toward base. Lettering goes round sides of stone and is picked out in modern red.

A fairly shelly, pale buff, oolitic limestone. Ooliths fairly well rounded, 1 mm diam., and composed of micrite and a little siderite. A large number of abraded, recrystallised shell fragments, mainly lamellibranch and brachiopod remains. Sparite cement and about 5% angular silt-grade quartz.

As might be expected, this milestone matches well the Barnack Free-stone, the quarries being less than 7 miles from Durobrivae. Specimens from the Lincolnshire Limestone from the quarries at Yarwell and Wansford were generally less shelly and with a greater iron content than both the milestone and the Barnack Stone proper. It is most likely, then, that the stone was quarried at Barnack.

25 DUROBRIVAE 2   R.I.B. 2235

   TL 119972   FLORIAN: 276

Roughly rectangular: 94x41x25 cm

Found 1785 in a field called Bridge Close in a ditch west of the north gate of the Roman town; now in the Museum of Archaeology and Ethnology, Cambridge[48]. It would have stood originally on Ermine Street (2c) and the stone surely measures the distance from the town (Durobrivae) on Ermine Street and not to the settlement at Castor to the north-east, as Wright suggests and wrongly terms Durobrivae. The find spot is ¾ mile south-west of Castor (not 1 mile as he suggests) and the milestone must in any case have been moved[49].

A pale buff or buff-grey, shelly oolite. Ooliths not too well rounded but fairly even in size, nearly 1 mm diam. and composed mainly of micrite; a few are structureless. A variety of recrystallised shell fragments present and a few intraclasts of similar rock. Sparite cement.

Very similar to 24 and will also have come from the Lincolnshire Limestone at Barnack.

26 UNKNOWN   R.I.B. 2239

                                         CRISPUS: 317-26

Roughly rectangular: 114x23x18 cm

Origin unknown; now in Museum of Archaeology and Ethnology, Cambridge[48]. Horsley saw it before 1732 in the Cotton Collection at Connington, Hunts., and thought it came from the Carlisle area but on geological evidence it is probably local. Split down the axis.

A pale buff, shelly oolite. Ooliths of micrite, variable in size, 0.7-1.3 mm. Shell fragments abundant and various, sometimes more than 1 mm in length. There are intraclasts of similar rock. Sparite cement; traces of siderite.

Very similar, geologically, to the stones from Durobrivae (Water Newton) and Girton and it is most probable that it had the same origin from the horizon of the Barnack Rag, Lincolnshire Limestone, at Barnack. It is quite likely that it originally stood on Ermine Street or perhaps King Street.

27 ANCASTER   R.I.B. 2242

   SK 986445   CONSTANTINE I: 307-37

Rectangular: 71x30.5x16.5 cm

        Found 1838 on west side of Roman road (2c) about ¼ mile north of
Ancaster and a little distance from the northern entrance to the station;
now in Grantham Museum. Lettering fairly neat and unweathered. Stone
broken or cut at base and rough at back.

        A good, pale buff, oolite freestone. Ooliths micritic and variable
in size, up to 1 mm; occasionally distorted. Some pisoliths. A few
intraclasts and shell fragments. Sparite cement.

        From the Castle Bytham-Clipsham area, northwards towards Lincoln,
the main building stone of the Lincolnshire Limestone is the Ancaster Freestone. This is the equi alent of the Ketton Freestone of further south,
below the Barnack Rag, and here it tends to be a little shelly45, 50.
Specimens from the Castle Quarry, Ancaster and New Quarry, Clipsham, were
both very similar to the milestone. The Ancaster Freestone does not
extend as far north as Lincoln, so it is most likely that this stone was
hewn at one of the several quarries local to Ancaster.

28 LINCOLN 1   R.I.B. 2240

   SK 973704   VALERIAN: 253-9

Rectangular: 71x35.5x20 cm

        Found in excavations in Sibthorpe Street or High Street, Lincoln, and
built into a window recess in St. Mary's Guild Hall; rediscovered and
removed in 1910; now in Lincoln Museum. The inscription, which is not
very neat, ends R(es)p(ublica) L(indensis). Stone rough at back from re-
use. It probable marked the first mile from Lincoln south-west along the
Fosse Way (5f).

        A buff oolite. Ooliths well rounded and variable in size, often
reaching 1.75mm diam. they are largely micritic with only occasional
shell fragments. Fine grained sparry calcite cement. Angular silt grade
quartz, about 20% by mode. Some limonite.

        In the Lincoln area the best stone is found in the Silver Bed of the
Lower Lincolnshire Limestone51. This bed in the Dean and Chapter Quarry,
just north of Lincoln, has not been examined but specimens from the
extensive section at Greetwell Hollow, east of the city, provide a reasonable match, though they tend to have less quartz than the milestone.

29 LINCOLN 2   R.I.B. 2241

   SK 976718   VICTORINUS: 268-70

Squared pillar: 223x41x41 cm

        Found 1897 opposite Lion and Snake Inn, Bailgate, Lincoln, at the
intersection of the main streets of the Roman town; now in Lincoln
Museum. The inscription, which is poor, ends a L(indo) S(egelocum)
m(ilia) / p(assuum) XIIII. The expansion of the S in the penultimate
line, as Segelocum, relies on the distance to Littleborough, which is
identified with Segelocum on the evidence of the Antonine Itinery (Iters
I and II). An earlier text has been chiselled off but traces can still be
seen at the top and below the last line of the secondary text.

        A light buff, oolitic limestone. Ooliths largely micritic, up to
1.2 mm diam. Some pisoliths, usually formed around shell fragments.
Together with other shell fragments they form about 60% by mode, the
remainder being microcrystalline calcite and occasional silty quartz.

        The Lincolnshire Limestone examined near Lincoln does not match this
milestone. The Hibaldstow Beds (Ancaster Freestone equivalent) here and
further north are often fissile and the ooliths frequently aggregate
together, so that they have seldom been used for building purposes45. I
would suggest that from the Lincolnshire Limestone examined, the milestone is most likely to have been quarried from the Ancaster Freestone in
the vicinity of Ancaster.

30 OWMBY   O.S. Map of Roman Britain (3rd ed., 1956)

SK 971870   UNINSCRIBED

Uninscribed Roman milestone seen before 1956 by some farm buildings at Owmby, Lincolnshire, on the east side of the Ermine Street (2d). I was unable to locate this stone on a brief examination of the area but it may still be there.

31 CAVE'S INN   J.R.S., liv (1964), 179

SP 535794   CONSTANTINE I: 307-37

Rectangular: 117x25.5x20 cm

Found in 1963 in a well a "score of yards from the Roman road Watling Street", (1f) at Cave's Inn (Tripontium), Shawell, Leicestershire; now at Rugby Fire Station. Lettering a little weathered but neat. Stone in three parts, the top being dowell jointed to the middle, smallest portion which is broken from the base.

A red, ferruginous, calcareous sandstone. Quartz grains vary from silt to medium sand grade, about 40% by mode. Calcite, detrital grains and cement, about 30%. Iron ores about 20%.

This stone is from the Northampton Sand (Inferior Oolite) which crops out through Northamptonshire and Rutland. According to Woodward[52] "this ferruginous and calcareous sandstone has been quarried in many places for building stone... It has been extensively used in Northamptonshire and Rutlandshire...at Heledon, Thorpe Mandeville, near Byfield, Eydon, Duston, Harleston, Desborough", etc. Duston, about 18 miles south-east from Cave's Inn, was an iron-working centre in Roman Britain[53] and it is reasonable to conjecture that this milestone was quarried there.

32 THURMASTON   R.I.B. 2244

SK 602076   HADRIAN: 119-20

Cylindrical: 96.5x53 cm diam.

Found 1771 on west side of Fosse Way (5f) near Thurmaston and about two modern miles from Leicester; now in the Jewry Wall Museum, Leicester. It is likely that the stone was found in situ, the find spot exceeding by 410 yards the distance of two Roman miles from the east gate of Ratae. Since the inscription ends "A RATIS / M II", it is unlikely that the distance was measured from Lincoln, as suggested by Wright in R.I.B.: rather it was measured from a point 410 yards outside the town boundary. It may be that the pomerium ran 410 yards outside the walls on the north side of the town, having been established before the town was built[54]. The stone and lettering are badly weathered but the lettering has been picked out in modern red paint.

A soft, poorly cemented, buff-orange orthoquartzite. Sorting good. A little clayey matrix and rare limonite grains.

A Triassic sandstone. There is little stone in the immediate neighbourhood of Leicester which can be used for building or inscribing but at Dane Hills the Upper Keuper Sandstone is sufficiently well cemented to have been quarried as a poor building stone[55]. Indeed, it was quarried for buildings in Roman Leicester[56].

33 SIX HILLS   R.I.B. 2245

SK 644208   UNDATED

Cylindrical: 56x38 cm diam.

Found 1851 beside Fosse Way (5f), near Six Hills, 11 miles north-east of Leicester; now in the Jewry Wall Museum. Upper part of a milestone, now badly weathered and rough. Inscription almost completely erased.

A soft, orange-red protoquartzite. Sorting good.

Very similar to the Thurmaston stone though there is a little more clayey, iron-rich matrix. Also Upper Keuper Sandstone from the Dane Hills.

34 WALL   R.I.B. 2246

   SK 097066   ? CLAUDIUS II: 268-70

Fragment of cylinder: 33x25.5 cm

 Found 1912 in excavations at Wall (Letocetum), by Watling Street (1g-h); now in Wall Museum. Lettering badly weathered and picked out in modern red paint.

 A buff or rusty-red, fine to medium grained, ferruginous sandstone (lithic greywacke). Sorting of quartz is poor; sphericity and rounding moderate. Clay minerals present are largely secondary, being altered feldspars. Iron ores are haematite and magnetite. Clay/silt matrix nearly 20% by mode.

 A sample of the normal building stone from Letocetum was examined and it is essentially similar to the milestone. Both are Keuper Sandstone which crops out locally in the Wall-Lichfield area.

35 BUXTON   R.I.B. 2243

   SK 06087330   UNDATED

Cylindrical: 61x33 cm diam.

 Found 1862 at Silverlands, Hardwicke Square, near the goods station in Buxton; now in Buxton Museum and Library. Lower portion of a roughly cylindrical stone from which the emperor's name has been lost but which measured the distance, "M P XI", (along road 710a) from the fort at Brough-on-Noe (Nauio). Lettering reasonably neat though the stone bears signs of re-use.

 A slightly buff, coarse, feldspathic quartzite. Quartz variable in size, moderate sphericity, 85% by mode. Feldspar, 6% by mode, is generally fairly fresh microcline. There are clay minerals, some iron ore and secondary silica cement.

 An Upper Namurian sandstone (Millstone Grit) from north-east Staffordshire or North Derbyshire. There are two or three feldspathic sandstones from which this milestone may come, most likely from the outcrops just west and south-west of Buxton. Geologically it is also possible that this stone came from the Bakewell-Eyam area to the east, but this seems unlikely on the archaeological evidence. It is unlikely, geologically, to have come from the Castleton-Brough area[57].

36 MOSTON POOL 1   R.I.B. 2247

   SJ 555275   UNDATED

Quadrangular, on base: about 140 cm high

 Found, with 37, in 1812, while draining Moston Pool, Stanton upon Hine Heath, Salop.; now lost. Found on a gravel bank beside the old road (6a) from Wroxeter to Chester and probably marked the 12th Roman mile from Viroconium.

37 MOSTON POOL 2   R.I.B. 2248

   SJ 555275   UNDATED

Quadrangular, on base: about 140 cm high

 Found, with 36, in 1812; now lost.

38 ALCESTER

SP 087578

J.R.S. lvi (1966), 220

CONSTANTINE I: 307-37

Rectangular: 74x34x20 cm

Found at Birch Abbey, Alcester, Warks., in 1965, incorporated into foundations of a late Roman building; now in Warwick Museum. It would have originally stood on the road (18b) which goes north, eventually to Letocetum. Broken at top and bottom.

A warm, buff-orange-brown, slightly oolitic, shelly limestone; a fair number of abraded shell fragments and patches of platy calcite. "Ooliths" are often rounded intraclasts ooliths, pisoliths, shell fragments and micrite-siderite mud. About 20% by mode of silt and sand grade quartz.

Describing the Pea Grit Series, Inferior Oolite, of the Cleeve Hill-Broadway area, L. Richardson writes that the pisolitic structure is usually absent in the main hill-mass; "in this region the Series is generally mainly replaced by the warm brown (orange of the commercial circles) 'Yellow Guiting Stone'."[58] It crops out from Temple Guiting, through Stanway, and by Fish Hill to Saintbury Hill. The best match obtained is from Stanway but it is possible that the stone was quarried nearer to Saintbury where the Roman road (18a) from Bourton on the Water, on the Fosse Way, leaves the Cotswold Edge and heads towards Alcester.

39 KEMPSEY

SO 848493

R.I.B. 2249

CONSTANTINE I: 307-37

Rectangular: 91x48x15 cm

Found shortly before 1818 in the garden of Parsonage Farm, now Court House, Kempsey, 4 miles south of Worcester, apparently in the foundations of a building; now in the Victoria Institute, Worcester. It would have originally stood on the road (180) from Gloucester to Worcester. Inscription badly weathered and picked out in modern red. Broken at top and bottom and split across face.

A very shelly, pale buff oolite. "Ooliths" are often recrystallised micrite, sometimes micrite pellets or rounded calcite fragments. A wide variety of shell fragments.

The nearest Jurassic oolites to Kempsey occur in the Inferior Oolite cropping out on Bredon Hill, about 10 miles to the south-east[59]. The Lower Freestone, which has been quarried about one mile north-east of Conderton, is a whitish, oolitic freestone containing shell fragments and it provides a fairly good match with this milestone.

If the Roman road (55) or trackway from Cirencester (Corinium) to Hales were continued northwards to Worcester or Droitwich, it would certainly cross or pass very near to Bredon Hill. Indeed, Roman buildings and other remains have been found in the vicinity, The origin of this milestone from Bredon Hill is, therefore, quite likely. It must be remembered, though, that geologically a match could probably also be found from the oolites of the main Cotswold Hill-mass.

40 KENCHESTER

SO 441429

R.I.B. 2250

NUMERIAN: 283-4

Roughly rectangular: 66x46x15 cm

Found 1796 on site of north wall of Kenchester (Magnis), 4½ miles west of Hereford; now in Hereford City Museum. Lettering fairly crude, now picked out in red. Inscription ends "R P C D", indicating erection by the civitas Dobunni. Top is rounded; top left hand corner and lower part broken.

A medium grained lithic sandstone or subgreywacke; weathered a dull red but a fresh surface is light grey with drak, red-brown, iron stained patches. Quartz, sorting and sphericity moderate; more than 70% by mode. Rock fragments are mudstones and siltstones, sometimes micaceous; nearly 20%. Iron ores largely in druses or patches; mainly secondary. Up to 2% garnet.

This rock is from the Old Red Sandstone and the amount of garnet contained suggests that it is of Downtonian Age[60], possibly the Holdgate Sandstones Group[61]. Both Downtonian and Dittonian rocks crop out close to Kenchester, for example on Dinmore and Garnons Hills, where they have been quarried in the past for building purposes[62]. The quarries are all now overgrown but it is quite probable that some of the sandstones were worked in Roman times. Two altars from Kenchester, now in Hereford Museum, are also, apparently, of the same stone.

41 PYLE   R.I.B. 2251

   SS 827822   VICTORINUS: 268-70

Rectangular: 117 cm high, 35.5 wide tapering to 20 cm at top

Found 1845 in a wall opposite Pyle Cottage, Pyle, Glamorgan, on the road (60c) from Cardiff to Neath (Nidum); now in the Royal Institution Museum, Swansea. Lettering crude.

A creamy or light grey, medium grained quartzite. Quartz usually fairly well rounded with good sphericity; nearly 90%. Inclusions of apatite common. Secondary clays and micas present and some secondary silicification.

This stone is from the Millstone Grit (Namurian) which crops out on the southern outcrop of the South Wales Coalfield, from Pencoed, through Aberkenfig towards Pyle. The Namurian here is often shaly but it contains hard, quartzitic bands which are often conglomeratic[63]. Much further east, towards Caerphilly and Risca, the narrow Namurian outcrop is similar[64], but further west, across the Gower Peninsula, the quartzitic bands are absent[65]. It is likely, therefore, that the milestone has a local origin. It is also possible that the stone was quarried on the northern outcrop of the coalfield between Kidwelly and Merthyr Tydfil, but this seems unlikely.

42 MARGAM   R.I.B. 2255

   SS 81628487   POSTUMUS: 258-68

Roughly rectangular: 63.5x25.5x23 cm

Found 1926 in Coal Brook, 320 yards south-south-east of East Lodge, Margam; now in the National Museum of Wales, Cardiff. Originally on road (60c) from Cardiff to Neath. Except for the face, it is very rough, perhaps from re-use.

Quartzite; weathered surface reddish-brown but a fresh surface is light grey or creamy. Quartz sorting, rounding and sphericity fairly good; some recrystallisation along grain contacts. Some grains have rhyolitic inclusions. Siltstone fragments with iron ores; about 5%. Matrix of clay, quartz mud and iron ores; about 5%.

Same stone as the milestones from Pyle (41) and Aberavon (44), and is local Millstone Grit.

43 PORT TALBOT   R.I.B. 2254

   SS 783873   MAXIMINUS DAIA: 309-13

Rectangular: 152 cm above ground x51x30.5 cm at top tapering to base

Found 1839 on course of Roman road (60c) on eastern outskirts of Port Talbot; now in Margam Abbey Museum. A post-Roman inscription is on the back. It has rounded edges; may once have been cylindrical. Upper right hand portion of stone lost since its discovery.

A fine-medium grained, orange-red subgreywacke with fine mottled appearance. Quartz sorting, sphericity and rounding poor to moderate; 70-75%. Some recrystallisation along grain contacts. Siltstone fragments present. Matrix of fine grained quartz, clay minerals and mica; about 15%. Some iron ore.

This stone is not a Rhaetic sandstone as suggested in R.I.B. but is Upper Carboniferous, Pennant Sandstone. A specimen of this latter, from near Margam, was examined. It was of orange-brown colour; the quartz was a little coarser than that in this milestone and clays, iron ores and matrix a little more abundant. However, specimens of Rhaetic sandstone from Pyle Church and the north side of Stormy Down were more mature and of finer grain size. They contained a similar proportion of siltstone fragments as this milestone but the clays, iron ores and amount of matrix were significantly less. The milestone probably came from the hills somewhere between Port Talbot, Margam and Maesteg, where the Pennant Sandstones crop out.

44 ABERAVON	R.I.B. 2251, 2256, 2252
SS 7589	i GORDIAN III: 238-44
	ii DIOCLETIAN & MAXIMIAN: 286-305
	iii LICINIUS: 308-24

Rectangular: 140x38x25.5 cm

Found in or before 1847 on west side of the channel called New Cut at Aberavon, Port Talbot; now in the possession of Mr. J. Blundell, Nottage Court, Porthcawl. Stone probably local and there is no reason to believe the tradition that it was imported as ballast from abroad. It would originally have stood on the Roman road (60c) from Cardiff to Neath. Primary inscription fairly neat but all three inscriptions are badly weathered, only the first being distinguishable with any ease. Lines 5 and 6 of the secondary inscription, "ET MARI / .VRE" seem to have been recut from "ET MAXI / MIANO". It is possible that "MARI / D VRE(NSIS)" was intended, the mason having cut Maridurum for Maridunum, since this latter is similar in meaning. In this case it is possible that reference to Carmarthen was intended, but K. Jackson maintains that the correct form for the name of Carmarthen was Moridunum[66]. It must be remembered, though, that the date of this re-cutting cannot be established and the possibility that it is post-Roman must not be overlooked.

A light grey or creamy, medium grained mature quartzite with occasional pebbles about 4 cm long. Quartz sphericity and rounding good, sorting fair. Some secondary silica cement and recrystallisation along grain contacts. Apatite and topaz occur as inclusions, rutile as a detrital mineral. Some clay minerals and traces of iron ores.

Virtually the same as 41 and will have the same origin from the Millstone Grit of the Pyle-Aberkenfig-Pencoed area.

45 MELIN CRYTHAN	R.I.B. 2257
SY 743962	DIOCLETIAN: 284-305

Irregular: 86x20 cm wide at top

Found in or before 1892 by the railway at Melin Crythan, 1½ miles south of Neath and by the road (60c) from Cardiff. It has been re-cut many times and is now a small stone pillar broken at the top and partly cut away at the left side.

A medium-light grey, medium grained subgreywacke; a fine mottled appearance on a fresh surface. Quartz sorting, rounding and sphericity poor to moderate. Fragments of siltstone present, often with clay minerals and iron ores. Secondary mica occurs with clay minerals; a little iron ore.

Same material as 43, from the local Pennant Sandstone, Upper Coal Measures, Carboniferous, from somewhere between Port Talbot, Margam and Maesteg.

46 LLANHAMLACH	R.I.B. 2258, 2259
SO 085271	i CONSTANTINIUS CHLORUS: 293-306
	ii CONSTANTINE II: 371-37 or
Squared stone: 21.5x28x29 cm	337-40

Found 1936, re-used in steps of the granary at Millbrook Farm, Llanhamlach, 2½ miles south-east of Brecon; now in Brecon Museum. It would have stood on the road (62a) from Abergavenny. Top, left side and base chiselled where it was cut from a rectangular milestone.

A fine grained, grey, calcareous sandstone weathered purplish-red for the outer 2 cm or so. Quartz sphericity and rounding poor to moderate. A few chert grains present. Calcite occurs as detrital grains and as cement. Calcareous siltstone fragments abundant. Chlorite, mica and clay minerals also present.

The stone is from the Senni Beds, (Lower Old Red Sandstone) which are usually purplish or green, micaceous, sometimes flaggy sandstones[67]. They crop out extensively along the south side of the Usk Valley from Trecastle to Brecon, from where the outcrop broadens eastward to include most of the Black Mountains. Cornstones (calcareous beds)[68] occur at many places and, though this stone is not a cornstone, the presence of the calcite suggests that it may have been associated with one. Such a cornstone crops out a mile or two east of Llanhamlach.

47 TRECASTLE HILL	R.I.B. 2260, 2261
SN 814312	i POSTUMUS: 258-68
	ii VICTORINUS: 268-70

Roughly rectangular

Found 1769, 2 feet below ground level near the site of an old inn on top of Trecastle Hill, Carmarthen. For more details, see R.I.B. Originally stood on road (62b) from Y Gaer (? Cicutio) to Llandovery (? Alabum); now lost. Probably damaged at sides and top.

According to R.I.B. it was of limestone. If so, it may well have been quarried from the Wenlock of Ludlow Series (Silurian) which crop out in the immediate vicinity.

48 DYNEVOR	R.I.B. 2262
SN 615225	TACITUS: 275-6

Seen in 1697 re-used as a corner stone in a small farmhouse at Dynevor (or Dinefwr), Llandeilo, Carmarthen; now lost. Original position not known but it presumably belonged to the road (623) from Llandovery to Carmarthen.

49 GWAENYSGOR	J.R.S., xlvii (1957), 230
SJ 075810	SEVERUS ALEXANDER: 231-5

Irregular: 119x30.5x29 cm

Found 1956 on the demolition of cottages near the church at Gwaenysgor, 1 mile south-east of Prestatyn, now (1968) in the possession of Mr. T. Pennant-Williams, Fforddlas, Prestatyn. The nearest Roman road (67a) is 4 miles to the south and it was presumably removed from there. Lettering shallow but reasonably neat. Probably originally rectangular but the right side has been chamfered away; broken at top and bottom.

A fine grained, greenish dolerite. Augite phenocrysts, slightly weathered and with secondary ilmenite, set in a ground mass of mainly plagioclase feldspar laths and crystals.

Dr. Neaverson states that "the nearest place it could have come from would be the Conway Valley or Shropshire"[69], and certainly there is no local dolerite[70]. Shropshire seems an unlikely source area bearing in mind the Romano-British geography. The dolerites of the Conway Valley are usually coarser than this milestone whereas the dolerites from further west, towards Bangor, are usually of this finer grained type. The most likely origin, therefore, is west of Caerhun, though it could have come from the Conway Valley.

50 BWLCH-Y-DDEUFAN	R.I.B. 2267
SH 7271	CONSTANTINE I: 307-37

Roughly rectangular: 56x25.5x9 cm max.

Found 1954 about 4 miles west of Caerhun Fort on north side of Roman road (67c) from Ro-wen to Llanfairfechan, near Bwlch-y-Ddeufan; now in the National Museum of Wales, Cardiff. Lettering very shallow; picked out recently in red paint. Broken at top; face back and sides are cleavage planes or natural joints; tapers slightly in width and depth.

A fine grained, compact, dark blue-grey igneous rock. A ground mass of mainly plagioclase feldspar but with some quartz, augite, biotite and traces of other minerals. The augite forms occasional phenocrysts which may show simple twinning.

An andesite or andesite-rhyolite. Locally to the find spot there are many outcrops of igneous rocks, tuffs and lava flows of Ordovician age and of variable composition in the andesite-rhyolite range. The milestone appears to be a natural fragment and is so hard that the inscription is very shallow. In this case it would seem reasonable to assume that the stone is quite local in origin from one of the nearby outcrops.

51 RHIWIAU-UCHAF 1   R.I.B. 2265

    SH 67917275   HADRIAN: 120-1

Cylindrical: 200x48 cm diam.

Found 1883 in a field called Caegwag on Rhiwiau-uchaf Farm, 2 miles south of Llanfairfechan, Caern., and nearly 7 miles west of Caerhun Fort (Canovium). It marked the 8th Roman mile on the road (67c) from Caerhun to Caernarvon (Segontium), "A KANOVIO / M P VIII". Now in the British Museum. Lettering shallow and neat. A good cylindrical stone: a slight indentation, about 1.25 cm across, in the centre of the top of the stone and a suggestion of criss-cross striations, indicate that it was probably turned.

A coarse, grey, pebbly quartzite. Quartz pebbles, reaching 4 cm in length, occur in bands, though quartz grains do not normally exceed 2 mm. Sorting poor, sphericity and rounding variable. Slight secondary silicification. Except for a little mud or silt grade quartz and rare limonite grains, there is no matrix.

There seems to be no outcrop to match this stone closer than the Cefn-y-Fedw Sandstone (Millstone Grit), which crops out from Llangollen and Minera, north to Hope Mountain and Halkyn Mountain[71]. This is a coarse, pebbly, quartzose sandstone. The pebbles, 3 cm or more in length on the south part of the outcrop, decrease in size northwards and die out around Mould. It is possible that the quarry site was as far north as Hope Mountain, near Ffrith, but it is more likely that the stone came from south of Minera.

It is most unlikely that an outcrop of Millstone Grit on Anglesey could have provided the stone for there "the rock is remarkably uniform, and is thick bedded, rather coarse and soft sandstone... Pebbles are very rare except at Bodorgan, where there is a coarse conglomeratic base"[72]. The Denbigh Grits, which crop out on the east side of the River Conway, are generally not coarse enough to be the origin of this stone and are also feldspathic[73].

52 RHIWIAU-UCHAF 2   R.I.B. 2266

    SH 67917275   SEPTIMIUS SEVERUS & CARACALLA:
                                        198-209

Oval section: 61 cm high, 38 cm maximum diam.

Found 1883 at Rhiwiau-uchaf within 10 yards of milestone 51; now in the British Museum. Lettering shallow and reasonably neat. Damaged at top and broken at base. May once have been cylindrical and have had earlier inscription removed.

A coarse, grey, pebbly quartzite, similar to 51, though the pebbles do not exceed 2.5 cm in length and there is some matrix of fine quartz grains with some iron ores and clay minerals.

Though there are differences in composition and texture between this milestone and 51, they are certainly from the same stratigraphic horizon and perhaps from the same quarry. This milestone also originated from the Cefn-y-Fedw Sandstone of the Llangollen-Minera area.

53 ABER   J.R.S., 1 (1960), 238[74]

   SH 66887338   POSTUMUS: 261-2

Irregular pillar: 117 cm high, 33 cm maximum width

     Found 1959, 500 yards south-east by east of Madryn Farm, Aber,
Caerns., and presumably marked the next Roman mile from Rhiwiau-
uchaf along the road (67c) from Canovium to Segontium; now in the
Museum of Welsh Antiquities, Bangor. It has an inverted inscription,
interpreted as a tri-partite Celtic name, on the left face. Lettering
poor and shallow. Broken at top with two inscribed flat faces adjoining;
back rounded; tapers towards base.

     A coarsely laminated stone; a greenish, finely grained chlorite-
mica schist. Flakes of white mica (muscovite) are seen on laminated
surfaces and there are quartz veins. Mica and chlorite are usually
secondary on feldspar.

     Schist of this kind can be found cropping out on Anglesey, in the
rocks of the Mona Complex. There are two or three different green schists
within the Mona Complex but the Gwna Green Schist is the most likely[75].
This crops out in several areas on Anglesey, including close to Parys
Mountain where copper was mined in Roman times, and on the Menai shore
opposite Bangor.

54 TY COCH   R.I.B. 2264

   SH 583696   CARACALLA: 212-7

Column: 122 cm high

     Probable milestone found about 1806 near Ty Coch, Pentir, about 2
miles south of Bangor and 7 miles north-east of the fort at Caernarvon;
now lost. Originally stood on road (67c) from Canovium to Segontium.
The stone, "about 4 feet in length", seems to have been broken from a
portion left underground. According to R.I.B. the final line of the
inscription is "IX" but R.C.A.M. Caern. II, lxiii, rejects this line on
the grounds that, though this would agree with the distance from
Caernarvon, the characters were not recorded until 40 years after the
disappearance of the stone.

55 LLYS DINORWIC   R.I.B. 2263

   SH 56656357   DECIUS: 249-51

Roughly rectangular: 140x33x20 cm

     Found in or before 1795 among hut remains at Cae'r-bythod, near
Llys Dinorwic, Caerns.; now in the Segontium Museum. Originally on
road (67c) from Caerhun to Segontium. Lettering shallow but fairly neat.
Back and right side rough and slightly rounded.

     A very fine grained, light green-grey rhyolite, weathering almost
white, with reddish-brown phenocrysts usually 1-2 mm in length. Thin
section shows these to be rounded phenocrysts of both quartz and orthoclase feldspar set in a very fine grained quartz-feldspar ground mass
which has a tendency to form flow structures around the phenocrysts.

     Greenly describes the Arvonian rocks of Arvon including a granite
and a quartz rhyolite, both of which might be the origin of this stone[76].
The former, perhaps the more likely, is essentially an aggregate of
quartz, red orthoclase and colourless albite, fining eastwards and
becoming a rhyolite at Port Dinorwic. The latter, the quartz-rhyolite,
crops out in two long oval tracts around Llyn Padarn and south-west
from Bangor. Both these rock types occur within a fairly small area
and could be described as local to the find spot.

56 RIBCHESTER 1   R.I.B. 2269

   SD 6535   ? GORDIAN III: 238-44

Cylindrical

     Found before 1732 when Horsley saw it in a garden at the west end of the town at Ribchester. He records some letters at the lower end which would probably have belonged to a text read in the reverse direction when the stone was in secondary use (cf. 64 and 85). It is possible that this was the text recorded by Dodsworthe and this is the same milestone as 58.

57 RIBCHESTER 2   R.I.B. 598[77]

   SD 6535   DECIUS: 249-51

Fragment of cylinder

    Found 1899 in one of the granaries in Ribchester Fort; now in Ribchester Museum. Though it has but 25.5 cm of circumference remaining and is only 18 cm high, the portion of inscription left is very similar to other Decius stones (see especially 13 and 58).

    A buff or buff-orange, iron rich sub-arkose with dark red-brown streaks and lines where the iron content is high. Sorting poor; sphericity and rounding poor to moderate. Feldspar mainly sodic plagioclase but microcline and altered orthoclase also present. Fibrous clay minerals and white mica occur. Iron ore is almost entirely limonite.

    A Namurian sandstone (Millstone Grit). These sandstones crop out from Longridge Fell, a few miles north of Ribchester, southwards to Blackburn and beyond, and also further north in the Forest of Bowland-Lancaster Fells area. However, from the old quarries visited in the Ribchester area, quite a good match was obtained from Longridge Fell, about four miles to the north. It is quite possible that this milestone was quarried there.

58 RIBCHESTER 3   R.I.B. 2268

   SD 6535   DECIUS: 249-51

Cylindrical

    Found in 1612 at Ribchester; now lost. (See also 56).

59 ASHTON 1   R.I.B. 2270

   SD 473580   PHILIP: 244-9

Irregular: 160x48x30.5 cm

    Found 1811, while ploughing "in the township of Ashton", 3 miles south of Lancaster; now in Lancaster Museum. It originally belonged to the Roman road (70d) from Wigan to Lancaster. Lettering small and crude; picked out in recent times in red paint. Line 4 of inscription interpreted by Wright as N(ostro) Watkin read it as MP III Rough shape, approximately cylindrical at edges and lower back; upper part and front squared suggesting it may have been cut down. Top rough and undressed.

    An orange-red, medium-coarse grained protoquartzite. Some authigenic silica. A few, fairly fresh feldspar grains, both microcline and sodic plagioclase. Clay minerals and iron ores occur, these latter sometimes around quartz grains.

    Probably Bunter Sandstone from further south than the find spot. According to Whitehead and Price[78], the Bunter Sandstone of the Preston district "may be well bedded or massive, is fine to medium-grained and varies in colour from white, grey or yellow to brown, red or purple;... Pebbles are rare". The Keuper Sandstone, however, is "almost invariably coarser than and more angular in grain than the underlying Bunter Sandstone. In colour it varies from grey to dull red, rarely being of such a bright colour as that characteristic of the Bunter". Another possibility would be the more local Millstone Grit, which is occasionally stained red, but this is usually described as a feldspathic sandstone[79].

    The Bunter Sandstone crops out in a fairly broad tract between Preston, Garstang and the southern end of Morecambe Bay. Exposure is not good but a Roman quarry somewhere in the area is quite probable.

60 ASTON 2   R.I.B. 2271

    SD 473580   DECIUS: 249-51

Roughly quadrangular: 158 cm high, 16 cm wide

    Found 1834 in the same field as 59, at Ashton, 3 miles south of
Lancaster; now in Lancaster Museum. Inscription shallow but reasonably
neat; recently picked out in red paint. A very "lumpy" stone with only
the face dressed flat for inscribing; probably an original loose rock.

    A light grey, feldspathic orthoquartzite with buff-orange stains.
Quartz poorly sorted; sphericity and rounding variable. Feldspar is
equally plagioclase and alkali types; often altered and replaced by clay
minerals and secondary mica or illite. Iron ores, mainly haematite and
magnetite, occur patchily.

    Millstone Grit, probably from the Lower Sandstone, which crops out
locally where not covered by drift. There is nothing sufficiently distinctive about the rock to suggest where it has come from but, as noted
above, the stone seems to be a naturally weathered block, in which case it
is probably very local in origin.

61 CATON   R.I.B. 2272

    SD 534646   HADRIAN: 119-38

Cylindrical; 240 cm above ground and 38 cm set in ground, 46 cm diam.;
rectangular base, 43x43x46 cm

    Found 1803 in bed of Artle Beck, near Caton, 4 miles east-north-east
of Lancaster, by the road to Burrow in Lonsdale (705); now in Lancaster
Museum. Inscription reasonably neat: marks at the beginning of the last
line have been variously interpreted as part of an ansate frame or an "L"
making the reading "L MP IIII" - "from Lancaster, 4 miles". Whichever
is correct, it is quite possible that line 5, the last line, or at least
the frame, was added later as it cuts very close to the bottom of line 4.
The milestone resembles 51, Rhiwiau-uchaf 1, but is not quite such a neat
cylinder. The tallest Roman milestone extant.

    A coarse, grey, feldspathic quartzite. Quartz grains often large
composite grains of igneous origin. Sorting poor; sphericity and rounding variable. Feldspar is mainly alkali feldspar; orthoclase and microcline. Secondary clays fairly common. Iron ore principally haematite.

    Millstone Grit from one of the Namurian Sandstones, probably from
one of the higher horizons, though a good match has not been found. The
sandstones crop out in many places in the Forest of Bowland and as far
north as the Lune Valley[79]. The origin, therefore, could be quite local.
Similar sandstones also crop out further south at Longridge, a few miles
north of Ribchester, and a possible match has been found there.

62 OVERTOWN

    SD 63017610   UNINSCRIBED

Roughly cylindrical: 60 cm above ground, approx. 45 cm diam.

    A milestone on line of Roman road (7c) from Ribchester to Low Borrow
Bridge, 150 yards south-east of Overtown Farm and ¾ mile east of the Roman
Fort at Burrow in Lonsdale. Still in place and marks the 60th mile from
Carlisle (Luguvalium) and so is the seventh south from the Middleton Milestone (63). May once have been cylindrical but is now somewhat flattened
on one side suggesting that it could once have been inscribed. Re-used
as a boundary stone and bears the modern lettering B / B / B and LECK.

    A buff-yellow, fine-medium grained protoquartzite. Sorting fair;
sphericity and rounding variable but generally poor. Feldspars rare but
clay minerals, kaolinite and illite, relatively abundant as secondary
products from feldspars. Heavy minerals include pyroxene and apatite.

    A sandstone from the Carboniferous Yoredale succession. One such
sandstone is very like another and there is nothing to suggest which
horizon this stone came from, or that it is not from the local outcrops.
Yoredales are exposed at places west of the River Lune, in the vicinity
of Kirkby Lonsdale and Burrow in Lonsdale. Exposure is better, however,

on Great Whernside and the hills behind Ingleton, a mile or two to the north-east[80].

63 MIDDLETON   <u>R.I.B.</u> **2283**

   SD 622858   UNDATED

Cylindrical: 163 cm above ground, 40 cm diam.

     Found 1836 in grounds of Hawking Hall, Middleton, 4 miles north of Kirkby Lonsdale and re-erected about 200 yards from the find spot. It stood on the road (70c) from Low Borrow Bridge to Burrow in Lonsdale. Apparently never inscribed with an emperor's name but bears only a distance, "M P LIII"; measured from Carlisle via Penrith and Sedburgh. The stone also has an inscription which commemorates its re-erection.

     A fairly typical Yoredale sandstone; a protoquartzite. Quartz sorting poor; sphericity and rounding poor to moderate. Feldspars rare but secondary clays present. Iron ores are limonite and haematite. Heavy minerals include pyroxene.

     As with the Overtown milestone, 62, there is nothing to suggest that this is not from one of the local Yoredale sandstones, from west of the River Lune near Burrow in Lonsdale or probably from the hills to the north-east of Ingleton.

64 CASTLEFORD 1   <u>R.I.B.</u> 2273, 2274

   SE 426251   i DECIUS & DECIUS: 250-1
                               ii GALLUS & VOLUSIAN: 251-3

Roughly cylindrical: 173x27 cm diam.

     Found about 1840 in Beancroft Road, Castleford (Lagentium), on the road (28b) from Tadcaster (Calcaria) to Doncaster (Danum); now in Leeds Museum. The second inscription was cut at the other end of the inverted milestone and ends "EB / MP XXII", (from York, 22 miles). Both inscriptions picked out in recent times in white paint. A dowel hole is in the centre of the Decius end and a broken metal pin at the other.

     A fine-medium grained, pale grey orthoquartzite, with slight iron staining. Sorting fair; some recrystallisation along grain contacts. Feldspars rare; some clay minerals replacing feldspars. A little limonite and magnetite.

     Origin uncertain but it seems to be a Carboniferous sandstone. There are no good petrographic descriptions of the local Carboniferous sediments but, though this is different from the other Castleford milestone, the literature available suggests that the stone is either a Coal Measures sandstone (Westphalian) and probably quite local[81], or perhaps a Namurian sandstone from several miles to the west or north-west[82]. What can be stated with certainty is that this is not a Triassic sandstone and therefore did not originate close to York or Aldborough.

65 CASTLEFORD 2   <u>R.I.B.</u> 2275

   SE 428257   FLORIAN: 276

Rough cylinder: 104x33 cm diam.

     Found about 1880 while drain laying in Carlton Street, Castleford, close to the line of the Roman road (28b) to Tadcaster; now in the Yorkshire Museum, York. It has a black encrustation; lettering deep, crude and weathered.

     A coarse, fairly soft, feldspathic quartzite. Basically a medium grey colour though individual grains vary from almost white to dark grey, sometimes with a dull red tinge. Sorting poor; sphericity and rounding variable. Feldspar, 5%, mainly alkali feldspar and often replaced by clay minerals, 4%. Iron ores are limonite, haematite and rare magnetite.

     A Carboniferous sandstone or grit, probably either Upper Namurian (Millstone Grit) or Lower Westphalian (Lower Coal Measures). The most likely is the Rough Rock which, south-west of Halifax (in the vicinity of

Slack (? Camulodunum)), is normally "coarse-grained and massive, though often crumbly where exposed owing to the decomposition of the large amount of feldspar it contains"[83]. It may, however, be from the Huddersfield White Rock or other horizons. We may reasonably conclude, though, that this milestone was quarried some miles to the west or north-west of Castleford.

66 DUEL CROSS   R.I.B. 2276

   SE 42706327   DECIUS: 249-51

Rough cylinder: 122x38 cm diam.

    Found 1776 at Duel Cross, 3 miles south of Aldborough (Isurium) on the road (8a) from York; now in the Lawson Museum, Aldborough. Lettering fairly neat though the stone appears to have been inscribed twice. Line 8, "XXG", all that remains of the primary text, has been interpreted as a mason's error for "XX", the distance from York, but as the find spot is 3 miles south of Aldborough and not 3 miles north, as given in R.I.B., this cannot be so.

    A coarse, poorly cemented, feldspathic quartzite; weathered orangy-red but patchy, buff orange inside. Sorting poor; sphericity and rounding variable but generally fair. 13% of quartz grains are composite; some recrystallisation along grain contacts. Feldspar, nearly 5%. is mainly orthoclase, but microcline and plagioclase also occur. There are some clay minerals. Iron ore, mainly haematite, is often associated with feldspars and clays.

    Red and orange Triassic sands and sandstones crop out east of the Pennines in a strip, averaging about 10 miles in width, from Nottingham in the south, through York, to Darlington and Middlesborough in the north. The pebbles, characteristic of the Bunter in the Midlands, die out north-wards and in Yorkshire there is no difference between the underlying Bunter and the Keuper[84]. The outcrop is often drift covered and not frequently exposed but in former times (earlier this century) the sand-stones were exposed in a small quarry just by Aldborough. The Triassic sandstone was used in the Roman town and the milestone would certainly have had a local origin.

67 ALDBOROUGH 1   R.I.B. 2277

   SE 406666   DECIUS: 249-51

Roughly oval: 41 cm high x33x23 cm

    Found 1924 in excavations 160 feet south of the north gate of Ald-borough; now in the Lawson Museum, Aldborough. Lettering very shallow but reasonably neat. It may once have been cylindrical but now roughly oval where face, and to a lesser extent sides, have been flattened. Broken at base and partly at back.

    A coarse, fairly soft, feldspathic quartzite with occasional small quartz pebbles. Red on outside but more patchy inside. Quartz poorly sorted; sphericity and rounding good. Some grains are composite; recrystallisation along grain contacts. Feldspar, 3.25%, is orthoclase and microcline with rare plagioclase. Clays, 6.6%, are platy and fibrous types. Iron ore, mainly haematite, often associated with clays and feldspars.

    A local Triassic sandstone, similar in origin to 66.

68 ALDBOROUGH 2   R.I.B. 2278

   SE 406664   DECIUS: 249-51

Rough cylinder: 69x35.5 cm diam.

    Found in or before 1852, near Aldborough, now in the Lawson Museum. Lettering and stone badly weathered. Upper part of milestone.

    A coarse, soft, red, feldspathic quartzite with occasional small quartz pebbles. Quartz variable in size; sphericity and rounding average fair; some composite grains. Feldspar, 5.2%, are orthoclase, microcline and plagioclase. Clay minerals occur. Iron ore mainly haematite, associated with clay minerals and feldspars.

A local Triassic sandstone similar in origin to 66 and 67.

69 PIERCEBRIDGE   R.I.B. 2293

NZ 207161   GALERIUS: 305-11

Rough cylinder: 140x40.5 cm

Found 1953 while gravel quarrying, 550 yards west of Dere Street (8c), near Piercebridge Fort (Magnis); now in Bowes Museum, Barnard Castle. Inscription shallow and partly erased; flattened sides and top. Lower half of stone more square than upper half and rough with ? chisel marks: it may well once have been the top, having had its inscription removed.

A fine grained, light grey orthoquartzite; weathers buff or buff-red on the surface. Quartz fine to medium sand grade; sphericity and rounding fairly good; a little recrystallisation and secondary silicific-ation. Feldspars rare but clay minerals more abundant (3%). Traces of haematite throughout. Rounded augite grains are conspicuous amongst heavy minerals.

Probably a Coal Measures sandstone (Westphalian) or perhaps Namurian sandstone; similar to Castleford 1 milestone (64) but most probably of different origin. It seems unlikely to have come from west of Piercebridge in the Cotherstone syncline area but rocks of similar age crop out extensively to the north of Piercebridge, towards Binchester (Vinovia). Its origin, therefore, is probably local. (Cf. milestones 73,74,75 and 76).

70 WILLINGTON 1   R.I.B. 2294

NZ 194354   GORDIAN III: 238-44

38 cm high x 38 cm wide

Found 1910 on line of Roman road (83) to the north-east of Willington, on the east side of Willington Burn. Re-discovered in 1941 at Laurel House, Willington; now a Cherry Tree House, Wolsingham[85]. I was unable to see and sample the stone. Part of a cylindrical stone; inscribed face rounded but lower and upper parts cut and back broken off.

According to R.I.B. it is of Millstone Grit. If so, it could well be of local origin.

71 WILLINGTON 2   Hutchinson, Durham, iii (1794), 318[86]

NZ 189340   ? UNINSCRIBED

Found ½ mile west of Willington shortly before Hutchinson's visit. He reports it was "squared down and defaced, to make the pillar of a shed for cattle", so it is unlikely it can be identified with Willington 1. Whether it belonged to Dere Street (8d) or to the side road (83) north-east from Willington, as did 70, is not clear.

72 LANCHESTER   R.I.B. 2295

NZ 160470   GORDIAN III: 238-44

Cylindrical

First seen 1783 used as a gate post on north side of modern road from Lanchester to Lanchester Fort (Longovicium), Durham; later seen at Greenwell Ford; now lost. Originally belonged to Dere Street (8d).

73 GRETA BRIDGE   R.I.B. 2279

NZ 08441328   GALLUS & VOLUSIAN: 251-3

Cylindrical: 200 cm x 38 cm diam.

Found in or before 1727 by the north-west angle of the fort at Greta Bridge, 20 yards south of Roman road (82) from Scotch Corner to Carlisle;

now at Rokeby House. Inscription reasonably neat. Stone somewhat broken at back and top; fracture below line 5.

A medium grained, orangy-yellow, ferruginous quartzite. Sorting fair; sphericity and rounding moderate; traces of recrystallisation along some grain contacts. Feldspars rare but a variety of clay minerals present. Iron ore, 11.5%, seems to be almost exclusively interstitial limonite.

More iron rich than the Stainmore milestones but otherwise the material is quite similar. Specimens obtained from the local Namurian sandstones provide a reasonable match and it seems safe to conclude that this stone originated locally in the Tees or Greta valleys.

74 STAINMORE 1	R.I.B. 2280
NY 949128	i FLORIAN: 276
	ii PROBUS: 276-82

Good cylinder: 81x35.5 cm diam.

Found 1924 in ditch beside the road (82) on Stainmore near Spittal, about 100 yards west of Vale House and 2½ miles west of Bowes (Lavatrae); now in the County Hall, Northallerton. Inscription a palimpsest, the second inscription using many letters of the first.

A medium grained, buff-grey, feldspathic sandstone. Quartz 80.5%; sorting moderate, sphericity and rounding variable but average moderate. Some quartz grains composite; a little authigenic silica. Feldspar, 4.2%, mainly orthoclase and often altered to clay minerals of various types; 10.8%. Iron ore mainly limonite.

A Carboniferous sandstone, and probably Upper Namurian since the Lower Coal Measures sandstones from near Barnard Castle contain carbonaceous matter. The Upper Namurian (Millstone Grit) from the Cotherstone syncline area[87] and Teesdale just north of Stainmore is variable in nature and there could well be several localities which provide a good match. The milestone, therefore, is probably local, though matching sandstones could perhaps also be found elsewhere, for example, within a few miles southwest of Catterick.

75 STAINMORE 2	R.I.B. 2281
NY 949128	CARUS: 282-3

Good cylinder: 66x30.5 cm diam.

Found 1924 with 74 on Stainmore; now in the County Hall, Northallerton. No "P" or numeral follows the "M" of the final line of the inscription, so the milestone might have been centrally inscribed, the number to be added, by cutting or painting, on the site. It has a dowel hole in the middle of the top and may originally have been intended as a column.

A medium grained, buff-grey, feldspathic sandstone. Quartz sorting moderate; sphericity and rounding fairly good. Some recrystallisation along grain contacts. Feldspars, 2%, often altered to clay minerals, 6.8%. Iron ore mainly limonite.

Though less feldspathic and clayey than Stainmore 1, it could well have the same quarry source from the local Upper Namurian (Millstone Grit).

76 STAINMORE 3	Ormerod, Leeds Phil. Soc. Procs.,
	i (1926-8), 259[88]
NY 949128	UNINSCRIBED

Cylindrical

Found 1927 within 100 yards of the spot where 74 and 75 were found on Stainmore; now in the County Hall, Northallerton. A crude cylinder, a little bigger than 74.

A medium grained, buff-grey, slightly feldspathic sandstone. Quartz sorting moderate; sphericity and rounding variable. Slight recrystallisation along grain contacts; a little authigenic silical present. Feldspar, 0.5%, altered to various clay minerals, 8.75%. Iron ore, 2.9%, is limonite and rare detrital magnetite.

Less feldspathic than 74 and 75 but otherwise very similar and is also local Upper Namurian sandstone (Millstone Grit). It probably originated from the same quarry, though perhaps from a different horizon.

77 TURNPIKE HOUSE: STAINMORE   R.I.B. 2282

    NY 906122   CARUS: 282-3

Cylindrical

Recorded in 1760 by Bishop Pococke and in 1776 by Hutchinson at the turnpike house, 5 miles west of Bowes on the road (82) to Carlisle; now lost.

78 HANGINGSHAW   R.I.B. 2284

    NY 682217   TWO PHILIPS: 244-6

Roughly rectangular: 61x33x30.5 cm

Found 1694 beside Roman road (82) from Scotch Corner to Carlisle at Hangingshaw, Long Marton, near Appleby; now in Tullie House Museum, Carlisle. Probably stood at the twenty-eighth mile from Carlisle. Lettering moderately neat though several lines continue onto the right side. It has the appearance of having been cut down from a larger stone for though it is roughly rectangular, it has slightly rounded sides. Broken at base.

A very slightly feldspathic quartzite, fine-medium grained and light grey in colour, weathering to buff on outside. Quartz sorting moderate to good; sphericity and rounding moderate or fair. Some recrystallisation along grain contacts. Feldspar, 0.8%, is orthoclase and albite. Various clay minerals occur, 5%. Iron ore, 0.3%, mainly limonite.

This stone is not from the local Penrith or St. Bees Sandstones though its actual origin is not clear. It is similar to the Castleford 1 and Piercebridge milestones and although it will undoubtedly have a different source from these, it is probably also Carboniferous. It is not a typical Yoredale sandstone but a match could probably be found somewhere in the Carboniferous to the east of the find spot, perhaps from the Namurian or Westphalian which is exposed on the west side of Stainmore[89]. Descriptions of the older rocks from the Cross Fell Inlier do not match[90]. Stone used for walling in the Hangingshaw area is a similar quartzite.

79 TEMPLE SOWERBY   Bruce, L.S., 747[91]

    NY 61962645   UNINSCRIBED

Roughly cylindrical: 137 cm high, about 30 cm diam.

Standing apparently in its original spot, about ½ mile south-east of Temple Sowerby, by the side of the modern road and on the line of the Roman road (82) from Kirkby Thore to Brougham. Perhaps once inscribed but now bears no trace of lettering. Irregular pillar, broken at top and tapering to base.

A medium-coarse grained, fairly well cemented protoquartzite. Quartz sorting fair; sphericity and rounding good. Some grains are compound, perhaps of metamorphic origin. Secondary quartz common, 15%, acting as a cement and forming overgrowths on many grains. Iron ore is mainly haematite which occasionally occurs as dense patches but often forms a thin skin of cement on quartz grains. It seems to have formed in two stages, one before and one after the deposition of the cement. Rare, altered feldspars present.

Penrith Sandstone: the formation crops out in a strip usually 3 or 4 miles wide, from Kirkby Stephen, north-west to Penrith and then north-north-west to within a few miles of Carlisle. At the extremes of its outcrop it is soft and poorly cemented but in the centre, from near Kirkby Thore to just north of Penrith, the extent of the silicification has ensured a fairly durable building stone[92]. It is most probable that this milestone, and those from Brougham, came from the hills just north or south of Brougham. There does not seem to be any recorded Roman quarry in the Penrith Sandstone, that at Crowdundle Beck (R.I.B. 998) being in St. Bees Sandstone.

80 BROUGHAM 1   J.R.S., lv (1965), 224

   NY 538293   POSTUMUS: 258-68

Rectangular: 76x51x30.5 cm

    Found 1964 while ploughing at Frenchfield, on north side of River Eamont at Brougham, beside the Roman road (82); now at Brougham Castle. The letters have been punched or pecked with a round chisel and not cut smooth. Inscription ends "R P C / CAR" - R(es)p(ublica) C(ivitatis) / Car(vetiorum). Probably marked the twentieth mile from Carlisle. Face and back dressed flat, sides rough.

    A medium-coarse grained, well cemented, red protoquartzite. Quartz well sorted; sphericity and rounding good. Some quartz grains composite. Much silica cement, 18.1%. Rare feldspars and clay minerals occur. Iron ore mainly secondary haematite having been deposited on the original quartz grains and after the formation of the silica cement.

    Penrith Sandstone quarried fairly locally, probably within 3 or 4 miles north or south-east of Brougham. (See 79).

81 BROUGHAM 2   R.I.B. 2285

   NY 536292   CONSTANTINE I: 307-37

Rectangular: 51 cm high, 20 cm wide

    Found 1602 near the fort at Brougham, near the confluence of Rivers Lowther and Eamont; now in Brougham Castle. Probably originally on the road (82) from Scotch Corner to Carlisle, but may have belonged to one of the other roads reaching Brougham. Lettering small but reasonably neat.

    A medium grained, fairly well cemented, red protoquartzite. Sorting of quartz moderate; sphericity and rounding fairly good. Secondary silica cements the grains, sometimes having formed after a deposit of haematite, though most haematite, the principle iron ore, 3.6%, formed after the cement. A little feldspar, 1%, mainly orthoclase. Some clay minerals, 1.9%.

    Penrith Sandstone quarried fairly locally, probably within 3 or 4 miles north or south-east of Brougham. (See 79).

82 OLD PENRITH   R.I.B. 2287

   NY 494385   VICTORINUS: 268-70

Roughly cylindrical

    Found in or before 1701 beside Roman road (7e) to Carlisle, near Old Penrith (Voreda); now lost. Apparently broken at base.

83 HESKET   R.I.B. 2288

   NY 4646   CONSTANTINE I: 307-37

66 cm high, 36 cm wide

    Found 1766 in Hesket on Roman road (7e) from Penrith to Carlisle. Seen in 1768 at Armathwaite by Gough; now lost.

84 SCALESCEUGH   R.I.B. 2284

   NY 448496   GORDIAN III: 238-44

Roughly cylindrical: 69 cm x 41 cm diam.

    Found 1915 north of farm buildings at Scalesceugh, near Carlton Hill,

5 miles south-east of Carlisle on road (7e) to Brougham; now in Tullie House Museum, Carlisle. Broken at top and bottom and left side flattened. Close examination of the stone shows that this left side is parallel to the bedding. As the face of a stone is normally dressed parallel to the bedding, it is likely that the left side of this stone was once inscribed.

A chocolate brown, fine grained subgreywacke. Dark lines parallel to the bedding are probably concentrations of magnetite. Sorting of quartz, 40%, fair; sphericity and rounding poor to moderate. Some grains of compound quartz. A high proportion, 27%, of ferruginous shale and silt-stone fragments. Feldspar, 0.5%, is mainly plagioclase: various clay minerals occur, 1.4%. Small flakes of white mica, muscovite, 0.9%. Iron ore, 9.9%, mainly haematite and magnetite. Augite is a common heavy mineral. Matrix, 9.2%, rich in iron and clay minerals.

A typical St. Bees Sandstone[93]. The formation crops out in a generally widening strip from Appleby, towards Carlisle, where it broadens to include Wigton, Carlisle and Brampton. The best building stone has been obtained in the Eden and Irthing valleys, east of Carlisle, though the formation has been quarried throughout its outcrop. Indeed, the Romans quarried St. Bees Sandstone at Crowdundle Beck (R.I.B. 998-1000) and south-west of Shawk (R.I.B. 1001-2) as well as at Wetheral (R.I.B. 1004-6) and along the River Gelt at Brampton (R.I.B. 1007-15). The St. Bees Sandstone is fairly uniform over much of its outcrop and so it is not possible to say with certainty where this milestone was quarried. However, of the specimens collected from the Roman quarry sites, the best match was obtained from Wetheral, 4 miles east-south-east of Carlisle, where the stone is deeper in colour, current bedded and, at lower horizons, shows the dark bands seen on this and other similar milestone specimens.

85 HARRABY BRIDGE   R.I.B. 2290, 2291, 2292

    NY 41325469   i CARAUSIUS: 286-93
                                        ii CONSTANTINE I: 306-7

Roughly rectangular: 188 cm high, 46 cm wide

Found 1894 in bed of River Petterill at Harraby Bridge, about 1 mile south of Carlisle (Luguvalium) on the road to Brougham (7e); now in Tullie House Museum. According to R.I.B. it has three inscriptions; a central text, presumed primary, chiselled away from the centre of the stone; a second text at the Carausius, inverted, broader end on the stone; and the final inscription to Constantine at the narrower end. It would be unusual, though, if the primary text were in the middle of the stone and it is more likely that this text was a continuation of the Constantine inscription or even a sub- or post-Roman addition to the stone. Carausius and Constantine inscriptions picked out in white paint. Face dressed flat; back rough.

A light grey, medium grained, slightly feldspathic quartzite, weathers to a buff colour. Sorting of quartz fair; sphericity and rounding moderate. Secondary silica, 4%, forms overgrowths on many grains. Feldspar, 2.3%, is mainly altered orthoclase. Clay minerals, 4.2%, various. Traces of limonite.

The origin of this milestone is not certain. The relatively low rounding and sphericity, and the small amount of secondary silica and iron ore make it certain that this is not Penrith Sandstone. As with the Hangingshaw milestone (78) it seems to be Carboniferous in origin, though the two most probably had different quarry sources. Lower Carboniferous rocks crop out 10 miles to the east but this milestone is not like the normal Yoredale sandstones that would be found there. Middle and Upper Carboniferous rocks are exposed about the same distance to the south-west of Carlisle and it may be that this milestone was quarried there.

86 OLD CARLISLE   R.I.B. 2286

    NY 260465   TWO PHILIPS: 247

Octagonal

Found in or before 1564 beside Roman road (75), near Old Carlisle, nearly 11 miles south-west of Carlisle. Removed to Rokeby, North Riding, where it was last seen by Gough in 1763.

87 OLD WALL   R.I.B. 2311

    NY 480616   DIOCLETIAN: 305-6

Rectangular: 71 cm high, 41 cm wide

     Found in or before 1816 at Old Wall, Irthington, on the Military Way
(86c) about 5 5/8 miles from Stanwix (Petriana), or 6 3/8 miles from
Carlisle (Luguvalium); now in Tullie House Museum. 132 cm high when
found but lower portion removed.

     A fine grained, chocolate coloured, slightly micaceous subgreywacke.
Fine black lines, which may be concentrations of magnetite, show small
scale current bedding. Sorting of quartz is moderate; sphericity and
rounding generally poor; some grains compound. A high proportion, 29.7%,
of shale and siltstone fragments. Rare feldspars and some clay minerals.
Muscovite present. Iron ore, 4.5%, is haematite and magnetite. Augite
occurs.

     St. Bees Sandstone. It is not possible to say with certainty where
it was quarried, but the nearest matching specimen obtained from a known
Roman quarry came from Wetheral (R.I.B. 1004-6), 4 miles east-south-east
of Carlisle. (See 84).

88 CARVORAN   R.I.B. 2310

    NY ?   CONSTANTINE: 306-7

Rectangular: 41x28x23 cm

     Found before 1716 near Carvoran (? Magna). Exact find spot not clear;
Heubner (C.I.L., vii, 1188) describes it as from "near Thirlwall Castle",
whereas Bruce (L.S., 326) writes that it was "found near Blenkinsop, which
is to the south of Magna, Carvoran". In the former case it would have
belonged to the Military Way (86b-c) or to the Stanegate (85a-b); in the
latter case it would have come from the Maiden Way (84), as Haverfield
suggests. Now in the Chapter Library, Durham. Lettering moderately
neat; picked out in black. Broken at base; split vertically from front
to back and rejoined.

     A fine grained, pale grey, micaceous protoquartzite. Sorting moderate;
sphericity and rounding fairly good. A few quartz grains are composite and
there is rare secondary silica. Feldspars are sparse, 0.8%, but clay
minerals abundant, 13.3%. Muscovite present, 1.5%, and the main iron ore,
5.7%, is limonite.

     Sandstone from the Yoredale facies (Carboniferous). It is well
matched by specimens from the Roman quarries at Coombe Crag (or Comb Crag)
(R.I.B. 1946-52), 4 miles west of Carvoran. The sandstone here is of the
Birdoswald Limestone Group which crops out along the Irthing valley to
Gilsland and beyond. So though it is possible that it was quarried at
Coombe Crag, it could also have been obtained from closer to Carvoran. In
any event, there is no reason to suppose that it was brought more than a
few miles.

89 FELL END   R.I.B. 2309

    NY 67466536   AURELIAN: 273-5

Rectangular: 66x33x20 cm

     Found 1932 re-used in a culvert on the Newcastle-Carlisle road near
Fell End Farm, more than ½ mile east of Carvoran; one of the series of
milestones from the Stanegate (85a); now in the Museum of Antiquities,
University of Newcastle. Originally probably rectangular; left side very
rough and broken away; top and right side bear tool marks from when it was
built into the culvert.

     A fine grained, chocolate coloured subgreywacke. Fine dark bedding
lines noticeable on the specimen are probably concentrations of magnetite.
Quartz sorting, sphericity and rounding are moderate; some quartz grains
compound. Fragments of ferruginous shale and siltstone fairly common, 20%.
Feldspars rather rare, 1%; various clay minerals a little more common,
2.4%. Muscovite rare, 0.3%. Iron ore, 1.5%, mainly haematite but
magnetite occurs. Augite is a noticeable heavy mineral.

St. Bees Sandstone. It is not possible to say with certainty where it was quarried but the closest match obtained from known Roman quarries came from Wetheral, 4 miles east-south-east of Carlisle. (See 84).

90 SMITH'S SHIELD   <u>R.I.B.</u> 2308

   NY 75676625   UNDATED

Cylindrical: now 61x48 cm diam., on base 28 cm high x48x48 cm

     Seen in or before 1725 standing on north side of Stanegate (85a) at Smith's Shield, Henshaw, almost one Roman mile west of Chesterholm Fort (Vindolanda). Base remains in position; upper parts of shaft re-used as gate posts 160 yards to the west, with no trace of the lettering. The part which in 1944 was being re-used as the north gate post (<u>R.I.B.</u> - half the column standing 127 cm above ground and 48 cm diam.) now seems to have disappeared; the other part (168 cm long but only 30.5 cm diam.) still, in 1968, lies dis-used near the gate. It had the remains of a primary text and the secondary, honorary inscription BON(O) / REI / PUBLIC AE / NATO. This second formula was used mainly for emperors of the fourth and fifth centuries[94].

     A medium grained, buff protoquartzite. Quartz sorting sphericity and rounding fair; a few compound grains of quartz present. A little recrystallisation along grain contacts. Feldspar rare, 0.6%, but clay minerals more abundant, 5.4%. Iron ores, 2.3%, limonite and magnetite. Heavy minerals include pyroxene and epidote.

     Sandstone from the Yoredale facies and not unlike the other local milestones which are of Yoredale sandstone. It could have the same origin as these others and probably local. (See 92).

91 CHESTERHOLM   Bruce, <u>L.S.</u>, 261[95]

   NY 77186649   UNINSCRIBED

Cylindrical: about 160 cm high, x55 cm diam.

     Still standing in its original spot on Stanegate (85a) about 110 yards east of Chesterholm Fort (Vindolanda) and one Roman mile east of the mile-stone at Smith's Shield. No inscription and hardly weathered enough to have had a complete inscription removed; a little damaged at top.

     A slightly feldspathic quartzite, fine-medium grained and buff-orange in colour. Quartz sorting moderate; sphericity and rounding poor to moderate; some quartz grains compound. Recrystallisation along some grain contacts. Feldspars, 2%, often weathered and altered to a variety of clay minerals, 2.3%. Main iron ore, 0.9%, is limonite.

     A sandstone from the Yoredale facies and probably local. It is not as coarse as 94 and 96 but in composition it is much the same as these and may well share a common origin. (See 92).

92 CRINDLEDYKES 1   R.I.B. 2299

   NY 78636697   SEVERUS ALEXANDER: 222-3

Rough cylinder: 112 cm high, 51 cm diam.

     Found 1885 at Crindledykes Farm, on north side of Stanegate (85a), one Roman mile east of the milestone at Chesterholm Fort (91); now in Chesters Museum. Found with four other milestones and two fragments (93-8). Inscription reasonably neat but shallow and weathered; it ends "MP XIIII", but records no <u>caput viae</u>.

     A light creamy-buff, fine-medium grained protoquartzite. Quartz sorting moderate; sphericity and rounding poor to moderate; some grains, 5.6%, compound. Some recrystallisation. Feldspars rather rare, 0.9%, but secondary clay minerals abundant, 10.7%. Iron ore, 1.8%, mainly limonite.

     A sandstone from the local Yoredale facies (Carboniferous). Sandstones occur at different horizons in a succession which includes many repetitions of the basic limestone-shale-sandstone-coal unit. Slightly more than ¼ mile south-west of the find spot, the sandstone above the Little Limestone

(Namurian)[96] was quarried by the Romans[97]. Specimens obtained here provide
a reasonable match with this milestone, but a better match was obtained
from the Roman quarry at Haltwhistle Burn[98]. This latter is now largely
overgrown but specimens were obtained which show that the stone quarried
here, the sandstone above the Three Yard Limestone (Visean)[96], is quite
variable in composition and could have supplied this milestone. The Roman
quarry at Bruton[97], a mile west of Chesters, is at a higher horizon,
probably the sandstone above the Little Limestone, and also provided a
reasonable match.

Clearly then, it is impossible to say with certainty where this and
other milestones of Yoredale sandstone were quarried. The best that can
be said is that in each case similar stone could be obtained locally, but
it is also possible that a milestone was quarried some distance from its
find spot.

93 CRINDLEDYKES 2   R.I.B. 2300

NY 78636697   PROBUS: 276-82

Rectangular: 71x41x23 cm, width tapers slightly to top

Found 1885 at Crindledykes Farm on Stanegate (85a) with 92 and others;
now in Chesters Museum.

A medium grained, creamy-yellow sandstone. It is now mounted high and
was inaccessible with the drill. On macroscopic examination it seems to be
very similar to or the same as 95. A sandstone from the Yoredale facies,
probably local in origin. (See 92).

94 CRINDLEDYKES 3   R.I.B. 2301

NY 78636697   i   CONSTANTIUS: 295-305
                                          ii   CONSTANTINE: 306-7
                                          iii MAXIMIANUS : 305-11
                                          iv   MAXIMINUS DAIA: 309-13

Rough cylinder: 160 cm high, 46 cm diam. at base tapering to 36 cm at top

Found 1885 at Crindledykes Farm on Stanegate (85a) with 92 and others;
now in Chesters Museum. Apparently inscribed four times, the earlier
inscriptions having been incorporated in the fourth: interpretation of
first and second texts and dates must, therefore, be conjectural.

A light, buff-grey, medium-coarse grained feldspathic quartzite.
Quartz sorting sphericity and rounding poor; some compound grains present.
A little recrystallisation of the quartz grains. Feldspar, 2.4%, mainly
orthoclase. Various clay minerals, 4.4%. The main iron ore is limonite.

A sandstone from the Yoredale facies and probably local origin.
Coarser and more quartzitic than 92 and no very good match was obtained from
the local proven Roman quarries. However the variation between this and 92
is probably within the range obtainable in some Yoredale sandstones and
they might have the same quarry source.

95 CRINDLEDYKES 4   R.I.B. 2303

NY 78636697   CONSTANTINE: 306-7

Cylindrical: 94 cm high x 30.5 cm diam.

Found 1885 at Crindledykes Farm on Stanegate (85a) with 92 and others;
now in Chesters Museum. Inscription fairly neat, on a rectangular panel
53 cm long and 35.5 cm wide.

A fine grained protoquartzite, light buff in colour. Quartz sorting
fair; sphericity and rounding moderate. Rare feldspar, 0.5%, and few
clay minerals, 1.5%, mainly kaolinite. Iron ore mainly limonite.

A sandstone from the Yoredale facies and probably local origin. More
quartzitic than 92, and fewer clay minerals, but this variation is quite
possible within one quarry and they may have the same source.

96 CRINDLEDYKES 5   R.I.B. 2302

  NY 78636697   CONSTANTINE I: 307-37

Rough cylinder: 117 cm high, 25 cm diam.

    Found 1885 at Crindledykes Farm on Stanegate (85a) with 92 and others;
now in Chesters Museum. The dimensions of this stone are considerably
less than many others of the Crindledykes group and it may well have been
re-inscribed on several occasions: face flattened.

    A light buff, medium-coarse grained protoquartzite. Quartz sorting
poor; sphericity and rounding fair; some compound grains. Rare feldspar,
0.2%, is orthoclase; clay minerals, 3.5%, mainly kaolinite. Iron ore,
1.1%, is limonite and magnetite.

    A sandstone from the Yoredale facies and probably local origin. Very
similar to 94 but less feldspathic. Noticeably coarser and more quartzitic
than 92, but the sandstones often have coarser horizons in a generally fine,
more massive sandstone and it is possible they have the same quarry source.

97 CRINDLEDYKES 6   R.I.B. 2304

  NY 78636697   UNDATED

Cylindrical: 28 cm high, 41 cm diam.

    Fragment found 1885 at Crindledykes Farm on Stanegate (85a) with 92
and others; once in the Summer House at Chesters, now lost. Top part of
stone on which only the beginning of the inscription, "IMP", remained.

98 CRINDLEDYKES 7   R.I.B. 2305

  NY 78636697   UNDATED

Cylinder on square base: 35.5 cm high, 35.5 cm wide

    Lower portion of stone found 1885 at Crindledykes Farm on Stanegate
(85a) with 92 and others; once in the Summer House at Chesters, now lost.
Inscription "L I" on one face of base has been variously interpreted but
Wright's suggestion (R.I.B.), that perhaps the letters "represent a builder's instruction for setting the stone", is the most plausible.

99 CORBRIDGE 1   R.I.B. 2296

  NY 982648   VICTORINUS: 268-70

    Found in or before 1760 at Corbridge (Corstopitum), where Bishop
Pococke saw it in the possession of the Rev. John Walton; now lost.

100 CORBRIDGE 2   R.I.B. 2297

  NY 981648   MAXIMINUS DAIA: 309-13

Originally roughly cylindrical: 112 cm high, 36 cm diam.

    Found 1867 on west side of the Roman town at Corbridge. It is not
clear whether it belonged to Dere Street (8d-e) or Stanegate (85a). Was
at Alnwick Castle, now in the Museum of Antiquities, University of
Newcastle. Inscribed twice but only traces of the primary inscription
remain beneath lettering on the secondary text. Back now trimmed flat
or split along a bedding plane. The face also flattened, probably when
the primary text was erased.

    A buff-grey, medium-coarse grained, very slightly feldspathic
quartzite. Quartz sorting poor; sphericity and rounding moderate.
Feldspar, 1.7%, mainly orthoclase, often altered to kaolinite and other
clay minerals, 4.23%. Iron ore, 2.9%, mainly limonite. A little
pyroxene present.

    Sandstone from the Yoredale facies and probably local origin. It
is similar to 96 and it is possible that they have the same quarry
source. (See 92).

101 COCKMOUNT HILL WOOD   J.C. Bruce, <u>Handbook to the Roman</u>
                                           <u>Wall</u>, 12th ed. by I.A. Richmond,
                                           (1966), 149

     NY 69466690   UNINSCRIBED

Good cylinder: about 140 cm above ground, 45 cm diam.

     Stone removed from Military Way (86b) and now forming the west post of
a gate at the west end of Cockmount Hill Wood, on the line of the Wall.
May well have been turned but no trace of lettering. It is broken at the
top and it is just possible that an inscribed portion has been lost off the
top, though the proportions would hardly allow for so much to have been
broken off.

     A coarse grained, orange-buff feldspathic sandstone. Quartz sorting
poor; sphericity and rounding variable, but generally poor; grains often
compound. Some recrystallisation. Feldspar, 4.1%, largely alkali feldspar
(microcline). Clay minerals mainly kaolinite, 7.9%. Iron ore, 0.6%,
limonite.

     Sandstone from the Yoredale facies. Noticeably coarser than any other
milestone but modal analysis essentially similar. No specimen as coarse as
this was obtained from the known Roman quarries but there is often
considerable variation laterally and vertically within the sandstones, so
this milestone is probably fairly local. (See 92).

102 CAWFIELDS 1   R.I.B. 2306

     NY 71776662   SEVERUS ALEXANDER: 222-3

Rough cylinder: 115 cm high, 46 cm diam.

     Found 1882 lying on south side of Military Way (86b) near milecastle
42, Cawfields, with 103 and 104; now in Chesters Museum. Lettering
shallow and worn but reasonably neat. Inscription ends "MP XVIII" and it
seems to measure the distance from Corstopitum along the Military Way, the
link road from Housesteads and the Stanegate.

     A light grey-buff, medium grained, very slightly feldspathic quartzite.
Quartz sorting, sphericity and rounding moderate; some grains are compound;
some recrystallisation of quartz. Feldspar, 0.8%, often altered to clay
minerals, 14%, principally kaolinite. Iron ore, 0.7%, mainly limonite.
Some pyroxene occurs.

     A sandstone from the Yoredale facies and probably local origin. In
grain size and modal analysis it is quite similar to 92.

103 CAWFIELDS 2   R.I.B. 2307

     NY 71776662   NUMERIAN: 282-3

Rough cylinder: 132 cm high, 46 cm diam.

     Found 1882 at Cawfields with 102, beside the Military Way (86b); now
in Chesters Museum. The stone is rough and the lettering almost worn away.
It is possible that there was an earlier text.

     A light grey-buff, medium-coarse grained, slightly feldspathic quart-
zite. Quartz sorting moderate; sphericity and rounding fair; a number of
compound quartz grains. Some recrystallisation along grain contacts.
Feldspar, 1.7%, mainly alkali feldspar including microcline. Clay minerals,
3.4%, various. Iron ore, 0.4%, is mainly limonite. Pyroxene and sphene
occur.

     A sandstone from the Yoredale facies and probably local origin. Modal
analysis more nearly like 94 and 98 than 92 and 102, though the presence of
pyroxene might suggest a common origin with these latter. However, it is
not impossible that they all came from the same quarry source.

104 CAWFIELDS 3   J.C. Bruce, Handbook to the Roman
                                   Wall, 12th ed., by I.A. Richmond
                                   (1966), 139-40[99]

    NY 71776662   UNINSCRIBED

Rough cylinder: about 120 cm high, 40 cm diam. tapering slightly

    Found 1882 beside the Military Way (86b) with 102, near Cawfields;
still at the find spot.

    A fine grained, buff-brown protoquartzite. Quartz sorting and
sphericity moderate; rounding poor; some compound grains. Some
recrystallisation along grain contacts. Feldspar, 0.7%, mainly plagioclase;
clay minerals, 8.9%, various. A few micaceous flakes and some siltstone
fragments, 1.9%. Tourmaline is a rare heavy mineral.

    A sandstone from the Yoredale facies and probably local origin. Darker
colour and of finer grain size than other milestones of Yoredale sandstone
and it perhaps had a different quarry source.

105 WELTON HALL   R.I.B. 2298

    NZ 06466824   CARACALLA: 213

Cylindrical

    Found 1813 lying on its face close against the Wall, nearly opposite
Welton Hall, east of milecastle 17. Probably came from the Military Way
(86a); now lost.

106 WATERFALLS   Dodds (ed.), Hist. Northl., xv
                                   (1940), 76[100]

    NY 91088082   UNINSCRIBED

Roughly cylindrical: about 150 cm high, 30 cm diam.

    Seen before 1702 on line of Roman road (8e) from Corbridge to
Risingham, when the modern road was lowered at the summit, 150 yards south
of the twelfth modern milestone from Corbridge. It marked the twelfth
Roman mile from Corstopitum, where the road changes direction. Re-erected
on the ridge behind Waterfalls Farm but in recent years it has fallen
down. Broken at top; the bottom 38 cm form a rectangular base.

    A fine-medium grained, buff-orange, very slightly feldspathic sand-
stone. Quartz sorting, sphericity and rounding poor; a few compound
grains. Some recrystallisation along grain contacts. Alkali feldspar,
0.8%. Kaolinite is the predominant clay mineral, 7.5%. Iron ore, 5.8%,
is mainly limonite.

    A sandstone from the Yoredale facies. In modal analysis it is not
unlike the milestones from the Wall area (see 92) but, though it is not
impossible that it shares a common origin with some of these milestones,
it can be matched fairly well with the local outcrops.

107 HIGH HOUSE   J.C. Bruce, L.S., 335[101]

    NY 896847   UNINSCRIBED

    Apparently uninscribed, first noted 1702 one mile south of Risingham
(Habitacum), near High House, Corsenside, on the Roman road (8e) from
Corbridge. It was two miles north of the milestone at Waterfalls Farm.
Now lost. According to Hunter it was "a pillar of about 8 feet in length,
which had stood by the wayside, but is now fallen".

108 INGLISTON   R.I.B. 2313[102]

    ? NT 123726   ANTONINUS PIUS: 140-4

Cylindrical: 64 cm high, 38 cm diam., originally more than 137 cm high

    Found before 1697 when recorded at Ingliston House, Kirkliston, and
believed to have come from Round Cairn, Newbridge. There is no authentic-

ated Roman road here but it is thought that there may have been one close
by, perhaps a continuation of Dere Street (8g) from Cramond to the Antonine Wall. Measured the distance from Newstead (Trimontium) on Dere Street,
though the actual figure is now missing. Now in the Museum of National
Antiquities, Edinburgh. Text on a raised panel defined by an ansate frame
(cf. 61 Caton, which has a sort of frame, and 95 Crindledykes 4, which has
its text on a panel): this, with mention of a military unit, make it unique
in Britain, perhaps in the empire. Whether or not the panel is contemporary
with the inscription is not clear. The initial "N" and "P" in lines 3 and
4 are a little detached from the main part of the inscription and are on a
portion of the milestone which is more deeply weathered than the main
inscribed part.

If the dating is correct, Cohors I Cugernorum should have been
stationed at Cramond during this period. There is no direct evidence for
this but we do know Trimontium was extended in the Antonine I period to take a detachment of the Twentieth legion from Chester[103]. This fort would probably
have had control over Cramond and the Antonine Wall area so its occurrence
as a caput viae is not unexpected.

A grey-brown protoquartzite. Quartz sorting good; sphericity and
rounding fair. There are siltstone and mudstone fragments which contain
iron ores and clay minerals. Various clay minerals, about 3%; rare white
mica. Iron ores are haematite and limonite.

The origin cannot be stated with certainty. Greensmith has described
the Lower Carboniferous oil-shale group sandstones which crop out locally
to Ingliston and have been quite extensively used for building stone[104].
These are frequently calcareous and contain a high proportion of clay
minerals so it is unlikely that this milestone came from here. The Lower
Old Red Sandstone of the Pentland Hills is usually described as coarse
and pebbly with conspicuous remains of older rocks and feldspars[105].
This seems to rule it out as a source for this milestone. The Upper Old
Red Sandstone has been quarried around Arthur's Seat. Descriptions are not
good[106] but these red and brown sandstones would be expected to contain
fragments of earlier Old Red sandstones and igneous rocks; so this again
is perhaps unlikely as a source.

MacGreggor and Eckford[107] described the Upper Old Red Sandstone of
Teviotdale and Tweedside as containing a little fairly inconspicuous
muscovite (white mica), wind rounded quartz grains, only a little feldspar
and no carbonaceous matter (as compared with the Lower Carboniferous of
the same area). It seems likely, therefore, that this milestone came from
the Upper Old Red Sandstone of the valley of Leader Water or upper
Teviotdale but an origin from the Upper Old Red Sandstone around Arthur's
Seat cannot be entirely ruled out.

109 BAR HILL   R.I.B. 2312

   NS 708760   ANTONINUS PIUS: 138-61

   Upper part of a rounded pillar with a rectangular panel, from the
Military Way (90) of the Antonine Wall, found in or before 1726 at Bar Hill;
now lost. Not clear whether it was a milestone as Stukeley, Wright and
others suggest, or a distance slab as Heubner and Macdonald believed[108].

110 MILECASTLE 64, HADRIAN'S WALL   J.R.S. lviii (1968), 209-10

   NY 418587   UNDATED

   Quadrant cut from a cylindrical milestone, 45.7 cm high with radius
17.8 cm and arc 25.5 cm: upper and lower portions cemented together.
Found 1962 on W. side of Milecastle 64 and presumed to have come from the
Military Way. Inscription MP. Now in Library of Army Apprentices College,
Hadrian's Camp, Carlisle. Described as 'red sandstone' but not sampled.

Schoolhouse,
Sandwick, Shetland

ABBREVIATIONS

Brit.   - _Britannia_

C.I.L. - _Corpus Inscriptionum Latinarum_

E.E.   - _Ephemeris Epigraphica_

J.R.S. - _Journal of Roman Studies_

L.S.   - J.C. Bruce, _Lepidarium Septentrionale_ (1875)

R.I.B. - R.G. Collingwood and R.P. Wright, _The Roman
          Inscriptions of Britain_, i (1965)

NOTES

1. <u>C.I.L.</u>, xii, 5519.
2. G. Walser, <u>Itinera Romana I</u>, Bern, (1967), 25.
3. <u>R.I.B.</u> 2265, described as Millstone Grit in <u>R.I.B.</u>
4. Numbers in brackets refer to milestones listed in the Appendix.
5. Mommsen in <u>C.I.L.</u>, viii (1881), 859-60.
6. C.E. Stevens, <u>English Historical Review</u>, lii (1937), 193-203.
7. "Late Roman" here refers specifically to the context of Roman Britain. Romano-British towns were given walls at a comparatively early date but Bitterne is peculiar in that its walls are at least mid-fourth century. In Gaul, where town walls are mainly post mid-third century, many inscribed stones were used in the walls. Bitterne is the only notable example in Britain.
8. S.H.A., <u>Vita Hadriani</u>, ii.
9. It has been suggested that milestones have been re-used as garden rollers but, so far as I am aware, no Roman milestone has been so employed.
10. F. Chayes, <u>Petrographic Modal Analysis</u> (New York, 1956); also M. Soloman, <u>Journal of Petrology</u>, IV (1963), 367-82.
11. A.L.F. Rivet, <u>Town and Country in Roman Britain</u> (revised ed. 1964), 64.
12. Other inscribed stones in Winchester Museum from the Belgic territory, on the evidence of macroscopic examination, are also Chilmark Stone. This suggested boundary between the Belgae and the Durotriges west of Rockbourne and Chilmark, is consistent with the evidence of the late Iron Age coins. A few coins of the Durotriges not unnaturally appear in the Belgic territory but the main occurrences are just west of Chilmark, a little nearer to Shaftesbury. (See D.F. Allen, <u>O.S. Map of Southern Britain in the Iron Age</u> (1962), introduction 19 ff, especially 30.
13. G.C. Boon, <u>Roman Silchester</u> (1957), 84, 92, 94, etc.
14. C.E. Stevens, <u>Procs. Somerset Arch. and Nat. Hist. Soc.</u>, xcvi (1952), 188-92.
15. I.D. Margary, <u>Roman roads in Britain</u> (2nd ed., 1967).
16. R.G. Collingwood in <u>V.C.H. Cornwall</u>, v (1925), 27-31.
17. For Clunch, see A.R. Warnes, <u>Building stones; their properties, decay and preservation</u> (1926), 59-60, and D. Purcell, <u>Cambridge Stone</u> (1967), 24-8. It weathers poorly in a smoky atmosphere but on an open Roman road the inscription would have been quite durable. The Corallian Coral Rock was also quarried by the Romans for building, a few miles north-north-east at Upware, but it is doubtful if this would have been suitable for inscribing (H.J. Green, <u>The Arch. Newsletter</u> vi, No. 12 (1960), 276-81).
18. C.E. Stevens, <u>op. cit.</u> (1937).
19. K.M. Kenyon, <u>Jewry Wall site, Leicester</u> (1948), 14.
20. <u>J.R.S.</u>, lvi (1966), 223.

21. Although the milestone from Aberavon (44) is probably
    local Millstone Grit, it is not impossible that it
    came from further north-west where there is a fairly
    narrow outcrop striking north-east from Kidwelly
    towards the Black Mountain. In this case it might
    relate to Moridunm, the civitas capital of the Demetae,
    as the second inscription (R.I.B. 2256) has sometimes
    been interpreted.
22. G. Simpson, Britons and the Roman Army (1964), 29.
23. J.R.S., lv (1965), 224, and milestone 80.
24. R.I.B. 1004-6.
25. I.A. Richmond in R.C.A.H.M. (Scotland), Roxburghshire,
    i (1956), 23-32.
26. I.D. Margary, op. cit.
27. H.G. Dines, S.C.A. Holmes and J.A. Robbie, The geology
    of the country around Chatham, Mem. Geol. Surv. G.B.
    (1954), 104-5, 149.
28. R.F. Jessup, Kent, (County Archaeological Series),
    (1930), 83.
29. I am grateful to A.L.F. Rivet for this observation.
30. H.J. Osborne White, The geology of the country near
    Brighton and Worthing, Mem. Geol. Surv. G.B. (1924), 7.
31. See also W. Lyon, Chronicles of Finchampstead, 5-7;
    Kempthorne, The Devil's Highway, Berks. Arch. Soc.
    (1901), 12.
32. For descriptions of Bath Stone see Anon., The Builder,
    lviii (1895), 273-8, 291-5, and for geological details,
    G.W. Green and D.T. Donovan, Bulletin Geol. Surv. G.B.,
    xxx (1969), 1-63.
33. G.C. Boon, op. cit.
34. C. Reid, The geology of the country around Salisbury,
    Mem. Geol. Surv. G.B. (1903), 8-15; A.R. Warnes, op.
    cit., 55-6.
35. M.A. Cotton and P.W. Gathercole, Excavations at
    Clausentum, Southampton, 1951-4 (1958), list the mile-
    stones found at Bitterne in Appendix I, but there seems
    to be some confusion in their record.
36. H.W. Bristow, The geology of the Isle of Wight, Mem.
    Geol. Surv. G.B. (2nd ed., 1889), 161, 251; F.J.
    North, Limestones (1930), 259.
37. E. Birley, J.R.S., lvi (1966), 228-31.
38. See also J.R.S., lvi (1966), 219.
39. Also R.H. Farrar, Procs. Dorset Nat. Hist. & Antiq.
    Field Club, lxxix (1957), 110-2.
40. H.B. Woodward, The Jurassic Rocks of Britain, iv
    (1894), 475-6.
41. R.M. Barton, An introduction to the geology of Cornwall
    (1964), 105-13; H. Dewey, British Regional Geology;
    South-West England (2nd ed., 1948), 32-3.
42. C.S. Exley and M. Stone, Present views on some aspects
    of the geology of Devon and Cornwall, Royal Geol. Soc.
    of Cornwall (1964), 131-84.
43. R.M. Barton, op. cit., 93 ff.
44. See R.G. Collingwood, J.R.S., xii (1922), 282 for a
    suggestion why this is so.
45. P.E. Kent, Trans. Leicester Philosophical and Literary
    Soc., lx (1966), 57-69; H.B. Woodward, op. cit., 171-4,
    204 ff.

46. D. Purcell, Cambridge Stone (1967), 29-34.
47. Ibid., 35-47.
48. Brit., ii (1971), 304.
49. For Durobrivae (Water Newton) and Castor see O.S. Map
     of Roman Britain (3rd ed., 1956), index, 22 and Fig. 3;
     and A.L.F. Rivet, op. cit., 146-7. The inscription
     is usually considered to end "M P I", but Miss J.
     Liversidge, Britain in the Roman Empire (1968), 38 (2),
     believes it to be "M P L" or "M P LI", suggesting that
     the distance recorded was measured from Lincoln.
50. D. Purcell, op. cit., 54-8.
51. Ibid., 70; W.D. Evans, Procs. Geologists Assoc., lxiii
     (1952), 316-35.
52. H.B. Woodward, op. cit., 475.
53. V.C.H. Northants., i, 197.
54. This suggestion finds support elsewhere. R.G. Goodchild,
     Berytus, ix (1949), 100, records a distance measured to
     a town boundary (pomerium) rather than a town centre.
55. C. Fox-Strangways, The geology of the country near
     Leicester, Mem. Geol. Surv. G.B. (1903).
56. K.M. Kenyon, Jewry Wall site, Leicester (1948), 14.
57. I am grateful to Professor F.W. Cope for this
     identification.
58. L. Richardson, The country around Morton in Marsh, Mem.
     Geol. Surv. G.B. (1929), 37.
59. Ibid., 45.
60. W.F. Fleet, Geological Mag., lxiii (1926), 505-16.
61. J.R.L. Allen and L.B. Tarlo, Geological Mag., c (1963),
     129-55.
62. B.B. Clarke, Trans. Woolhope Naturalists Field Club,
     xxxiii (1950), 97-111; and B.B. Clarke, ibid., 222-37.
63. A. Strahan and T.C. Cantrill, The geology of the South
     Wales Coalfield vi, Bridgend, Mem. Geol. Surv. G.B.
     (1904).
64. A.W. Woodland and W.B. Evans, The geology of the South
     Wales Coalfield iv, Pontypridd and Maesteg (3rd ed.,
     1964).
65. A. Strahan, et al, The geology of the South Wales
     Coalfield viii, Swansea (1907); A.E. Trueman,
     Procs. Geologists Assoc., xxxiv (1924), 283-308.
66. K. Jackson, J.R.S., xxxviii (1948), 53-8.
67. F.J. North, Brycheiniog, i (1955), 39; J. Pringle
     and T.N. George, British Regional Geology: South Wales
     (2nd ed., 1948), 49.
68. F.J. North, Limestones (1930), 113-4.

69. In T. Pennant Williams, Flintshire Hist. Soc. Pubns.,
     xvii (1957), 90.
70. E. Neaverson, Procs. Liverpool Geol. Soc., xv (1930),
     172-212.
71. G.H. Morton, Procs. Liverpool Geol. Soc., iii (1878),
     152-205, 299-325, 371-428; C.B. Wedd and W.B.R. King,
     The geology of the country around Flint, Hawarden and
     Caergwle, Mem. Geol. Surv. G.B. (1924).

72. E. Greenly, The geology of Anglesey, ii (1919), 660.
73. P.G.H. Boswell, Procs. Geologists Assoc., xlvi (1935), 152-86.
74. See also J.R.S., lii (1962), 195.
75. E. Greenly, op. cit., 67 ff.
76. E. Greenly, Quarterly Journal Geological Soc., c (1944), 269-87.
77. See also E. Birley, J.R.S., lvi (1966), 230.
78. In D. Price, et al., Geology of the country around Preston, Mem. Geol. Surv. G.B. (1963), 76-7.
79. F. Moseley, Quarterly Journal Geological Soc., cix (1953), 423-54.
80. K.C. Dunham, et al., Procs. Yorkshire Geol. Soc., xxix, pt. 2 (1953), 77-115.
81. G.H. Mitchell, et al. Geology of the country around Barnsley, Mem. Geol. Surv. G.B. (1947).
82. W. Edwards, G.H. Mitchell and T.H. Whitehead, Geology of the district north and east of Leeds, Mem. Geol. Surv. G.B. (1950); R.G.S. Hudson, et al., Procs. Geologists Assoc., xlix (1938), 295-352; D.A. Wray, et al., The geology of the country around Huddersfield and Halifax, Mem. Geol. Surv. G.B. (1930), 15-25.
83. D.A. Wray, et al., op. cit., 22.
84. W. Edwards and F.M. Trotter, British Regional Geology: the Pennines and adjacent areas (1954), 64; F. Smithson, Procs. Geologists Assoc., xlii (1931), 125-56.
85. Brit., i (1970), 315.
86. See also with R.I.B. 2294.
87. H.G. Reading, Quarterly Journal Geological Soc., cxiii (1957), 27-56.
88. See also with R.I.B. 2280.
89. T.D. Ford, Geological Mag., xcii (1955), 218-30.
90. F.W. Shotton and F.M. Trotter, Procs. Geologists Assoc., xivii (1936), 376-87.
91. See also R.C.H.M. Westmorland, 226; and P. Ross, Trans. Cumbd. and Westmd. Ant. and Arch. Soc. xviii (1918), 219-22.
92. H.C. Versey, Geology of the Appleby district, (3rd ed., 1951).
93. E.E.L. Dixon, et al., The geology of the Carlisle, Longtown and Silloth district, Mem. Geol. Surv. G.B. (1926), 23.
94. For the use of this formula see, E. Birley, Trans. Cumbd. and Westmd. Ant. and Arch. Soc. (NS), lviii (1958), 90.
95. See also Horsley, Brit. Rom. Northl., lix; Hodgson, Hist. Northl. II, iii, 201; Procs. Soc. Antiq. Newcastle, (ser. 3), v, 184.
96. G.A.L. Johnson, Procs. Yorkshire Geol. Soc., xxxii (1959), 83-130 and map.
97. O.S. Map of Hadrian's Wall (1964).
98. R.I.B. 1680.
99. See also with R.I.B. 2306.
100. O.S. Map of Roman Britain (3rd ed., 1956), index, 43.
101. See also Hunter, Phil. Trans., xxiii (29 May 1702), 278, 1132; Hunter, supra abridged in Phil. Trans., iv (1809), 667; Wallis, Hist. Northl., ii, 58; Dodds (ed.), Hist. Northl., xv (1940), 76, n 5.

102. See E. Birley, J.R.S., lvi (1966), 230, for the
     dating of this stone and Brit., iv (1973), 336,
     for recent information concerning the find spot
     and original size.
103. I.A. Richmond, R.C.A.H.M. Scotland, Roxburghshire,
     i (1956), 23-32.
104. J.T. Greensmith, Procs. Geologists Assoc., lxxii
     (1961), 49-71.
105. M. MacGreggor and A.G. MacGreggor, British Regional
     Geology: The Midland Valley of Scotland (2nd ed.,
     1948), 20-1.
106. B.N. Peach, et al., The geology of the neighbourhood
     of Edinburgh, Mem. Geol. Surv. G.B. (2nd ed., 1910),
     42-6.
107. A.G. MacGreggor and R.J.A. Eckford, Trans. Edin.
     Geol. Soc., xiv, pt. 2 (1948), 230-52.
108. M. Gichon and B.H. Isaac have recently (Israel
     Exploration Journal, xxiv (1974), 117-23) compared
     this stone with a Flavian inscription from Jerusalem,
     both of which they consider to be commemorative pillars
     rather than milestones.

www.ingramcontent.com/pod-product-compliance
Lightning Source LLC
Chambersburg PA
CBHW051309270326
41929CB00029B/3470